Electronic Warfare Signal Processing

For a listing of recent titles in the
Artech House Electronic Warfare Library,
turn to the back of this book.

Electronic Warfare Signal Processing

James Genova

ARTECH HOUSE
BOSTON | LONDON
artechhouse.com

Library of Congress Cataloging-in-Publication Data
A catalog record for this book is available from the U.S. Library of Congress

British Library Cataloguing in Publication Data
A catalog record for this book is available from the British Library.

ISBN-13: 978-1-63081-460-1

Cover design by John Gomes

© 2018 Artech House
685 Canton St.
Norwood, MA

All rights reserved. Printed and bound in the United States of America. No part of this book may be reproduced or utilized in any form or by any means, electronic or mechanical, including photocopying, recording, or by any information storage and retrieval system, without permission in writing from the publisher.

All terms mentioned in this book that are known to be trademarks or service marks have been appropriately capitalized. Artech House cannot attest to the accuracy of this information. Use of a term in this book should not be regarded as affecting the validity of any trademark or service mark.

10 9 8 7 6 5 4 3 2 1

To my wife, Libba

Contents

	Preface	**xi**
1	**Introduction to Modern EW**	**1**
1.1	Evolution of Naval EW	2
1.2	Terminology and Model Scenarios	15
1.3	Probability of Raid Annihilation	22
1.4	Sample Strategies	26
	References	29
2	**Pulsed Doppler Radar Basics**	**31**
2.1	Electromagnetic Pulse	33
2.2	Dynamic Range and Gain Control	44
2.3	Coherent Gain and Noncoherent Gain	56
2.4	Antenna	62
2.5	Doppler Effect	68
	References	76

3	**LPI Radar and EA Model**	**79**
3.1	ASM Model	81
3.2	Radar Range Equations and Burn Through	94
3.3	Range Doppler Map and Imaging	96
3.4	Target Scatter Model	99
3.5	Repeater EA Model and the DRFM	105
3.6	Summary of Model	108
3.7	Detection versus Classification and EP	112
	References	114

4	**Extended Target EP Signal Processing**	**115**
4.1	Target Classification: False Targets	117
4.2	Target Classification: Decoys	140
4.3	Target Classification: Chaff	142
4.4	Dual Coherent Source EA	146
	References	160

5	**LPI Radar EP Waveforms**	**163**
5.1	Coded Waveforms EP	165
5.2	Stepped Waveforms EP	173
5.3	Probe Waveforms EP	181
	References	185

6	**Multiple Receiver EP Signal Processing**	**187**
6.1	Dual-Coherent Source EP Approximation	189
6.2	ASM STAP Processing	199
6.3	Cover Jamming EP	207
6.4	Summary	220
	References	222

7	**Adaptive EW**	**223**
7.1	Overview	227
7.2	Fundamentals of LLR	235
7.3	EW Specifics	245

| 7.4 | Summary and Conclusions | 250 |
| | References | 252 |

About the Author   **253**

Index   **255**

Preface

The primary tactical function of the autonomous threat sensor is localization in the tracking mode after target detection in the surveillance mode while operating in the presence of jamming. Many common electronic attack (EA) techniques exploit flaws in the sensor to corrupt or degrade the sensor detection or tracking capability. To develop these EA techniques requires intelligence agencies to collect a sample of the actual threat sensor so the EW engineer can discover the particular sensor hardware flaw.

For example, in studying a particular threat sensor, a specific EA technique would be developed that can capture the sensor tracking gates. An EA waveform then introduces a low duty cycle false target (countdown technique) making the tracking loop unstable without the sensor realizing it is interrogating a false target. Other examples of classical EA techniques that attack the detection capability include the use of chaff to seduce the sensor, or the use of noise jamming to blind the sensor and prevent or delay target detection.

The electronic warfare (EW) battle is now an information battle. The classical radar anti-ship missile (ASM) seeker of the past is replaced with radio-frequency (RF) hardware that collects low power coherent data via multiple receiver channels for rapid and effective digital signal processing (DSP). With present radar technologies the modern autonomous threat sensor can readily detect and accurately locate multiple targets even in the presence of jamming. These potential targets are continually subjected to advanced DSP techniques for target feature analysis.

This information gathering capability of the modern radar sensor coupled with high-speed digital signal processors leads to a significant shift in emphasis of EW. The emphasis for the threat radar engineer is now on electronic protection (EP), in the sense of target identification and classification via target feature extraction and analysis. With the implementation of these practical digital EP techniques the modern radar sensor can quickly and reliably identify the correct target while collecting very precise guidance measurements. This improved sensor capability renders current EA techniques obsolete and affords autonomous threats an unprecedented tactical advantage.

The main purpose of this book is to describe the common techniques currently used by the modern autonomous radar sensor to classify targets. This book explains the current DSP algorithms that are most useful for an autonomous sensor to quickly and accurately make target classification decisions during its engagement while deliberately countering standard EA weapons. The goal is to give the reader an intuitive understanding of the basic EP technical issues. Throughout the book a simple physics-based mathematical model is developed and described to guide the reader's understanding.

This change in emphasis applies to all of modern EW, but is particularly evident in naval EW. I have been an active participant in the development and testing of EW systems for over 40 years. While I have worked on airborne and ground-based EW systems, my primary experience has been naval systems. I was involved in the development and testing of various naval onboard and off-board jamming systems. My most recent efforts have been in the development of hardware simulators for the EW research of threat radar sensors for ASMs. I led the programs to implement the first several coherent radar seeker hardware simulators for the U.S. Navy. I am, thus, uniquely qualified to explain these ASM capabilities. For these reasons, and because most EW texts focus on airborne EW, particular points throughout this text are demonstrated by the example of a radar guided autonomous ASM attacking a ship.

The naval fleet is a means for a nation to project visible power into a geographical area controlled by an enemy force. A typical fleet consists of at least one aircraft carrier together with an assortment of escort ships. While the escort ships possess the capability to launch weapons a considerable distance, their primary role is to protect the carrier as a mobile base for aircraft.

As a counter to this power projection, various nations have developed and continuously improved the ASM threat for more than 60 years. With improved navigation systems the modern ASM can fly multiple way-points over a considerable distance to the fleet. With improved sensors the ASM can seek out the ship target and autonomously guide to its intended target.

The most reliable and effective defense against the ASM is the variety of hard-kill weapons, for example anti-missile missiles, CIWS, and the high-energy laser beam. The goal of these weapons is to physically damage or destroy the ASM platform or sensor. However, a wave attack of modern ASM threats against a naval fleet presents a formidable challenge to fleet defenses. Such an attack is expected to overwhelm existing hard-kill assets. This possibility necessitates an effective fleet EA or soft-kill weapon defense capability as a complimentary adjunct to hard-kill defense.

Since an autonomous threat wave attack can overwhelm kinetic weapons and the sensor utilizing these DSP techniques can counter most classical EAs, defensive EW must evolve into a reliable and viable weapon system. To develop this defensive capability the EW engineer must understand these DSP techniques.

Understanding the material contained in this book is essential in focusing the efforts of EA development engineers to make the ship target appear less like a ship and/or to make a decoy appear more like a ship. Understanding this material is essential to guide intelligence personnel to focus their information gathering efforts to correctly assess the ASM sensor EP capabilities.

This book is divided into two parts. In the first part, the model details are developed and described together with a presentation of the fundamentals of radar and EW. Relevant examples are presented for use later in the book. There are many excellent books on EW such as *Electronic Warfare in the Information Age* by D. Curtis Schleher and there are many excellent books on the radar such as *High Resolution Radar* by Donald R. Wehner and *Microwave Radar* by Roger J. Sullivan. While the material herein is self-contained the reader should consult these or similar works for additional details of EW and radar processing fundamentals.

Chapter 1 describes the quantitative approach to naval warfare, using the concepts of probability of raid annihilation to relate the general, complex engagement to the engagement of a single ASM against a single ship. Definitions of many useful terms describing the naval engagement are given. Several example EW strategies are introduced and described for use throughout this book.

Chapter 2 contains the mathematical definitions of particular aspects of the ASM low probability of intercept radar sensor needed for understanding the later material. The basic concepts of pulsed radar are introduced with enough mathematical rigor to give the reader an intuitive understanding of the sensor. The particular aspects of the ASM sensor useful for later understanding are described. The book *Detecting and Classifying Low Probability of Intercept Radar* by Dr. Phillip Pace is an excellent reference for this chapter.

Chapter 3 brings the prior material together to present a physics-based mathematical model of the ASM radar sensor for a point target with and without the presence of EA. This chapter introduces the concepts of the digital RF memory (DRFM), repeater EA, cover noise EA, and burn through. In addition, this chapter introduces the basic understanding of modern EP and target classification techniques.

In the second part of this book specific modern algorithm approaches to the EP decision process are presented. Each topic demonstrates improvements to the ASM radar sensor that must be understood by the EW engineer to successfully develop effective EA techniques and by the intelligence engineer to properly assess the threat potential.

Chapter 4 describes EP techniques based on the ship being an extended target rather than a point target. Concepts of classical EA and angle deception for modern naval defense are described. Specific topics include a description of decoys and chaff as well as repeater EA, cover noise jamming, and the use of dual coherent source jamming to mimic an extended target.

In this chapter, it is shown how the modern digital processor makes it feasible to fully exploit the statistical properties of various targets. Using these simple algorithms the modern ASM radar seeker can survey the scene and quickly and autonomously distinguish decoys and chaff from ship targets. The EA engineer must take particular care to mimic these ship target features or to exploit these differences in a novel manner. Some proven basic EA concepts for synthesizing a complex false target are described.

Chapter 5 describes the special topic of ASM waveform design. Previously, waveforms were designed for improved detection and improved tracking parameter estimation. With modern technologies it is now feasible to design waveforms for enhanced target classification parameter estimation without sacrificing detection or tracking capability. The proactive design of waveforms for enhanced target classification is a new sensor capability. Waveforms are described that deliberately probe the target for enhanced feature information to successfully mitigate EA and decoy effectiveness.

Chapter 6 is a particularly important chapter. A major technological advance in the 1970s was the advent of monopulse angle estimation. Until recently the second receiver in the ASM radar sensor was only used for monopulse angle estimation. Now, with the availability of fast digital processors, ASM sensors are able to fully exploit the capability of these multiple receiver radar sensors. This capability totally mitigates the usefulness of cover noise jamming while providing much improved target detection and tracking capability.

In Chapter 7, the characteristics of a sensor fusion and feedback control algorithm are described. Such an algorithm is required to make the fleet

EW system a useful defensive weapons system. This algorithm illustrates the essential features of real-time EW effectiveness assessment or a posteriori probabilities of effectiveness required to raise EA to a useful level as a viable weapon option alongside kinetic weapons. Similar information fusion algorithms must be used by the ASM general processor to optimally combine the results of the multiple EP algorithms presented in the previous chapters.

    I appreciate the support of the many people who made this book possible. In particular, Mr. Al DiMattesa (retired) of the Naval Research Laboratory has been a mentor and supporter of my efforts in this area since 1978. And I want to thank Professor Phillip Pace (Naval Postgraduate School), who inspired the effort necessary to author this book.

1

Introduction to Modern EW

An autonomous threat such as an anti-ship missile (ASM) uses its sensors and subsequent processing to detect and select the correct target and to accurately measure guidance information for that target. With this information the ASM reliably delivers its ordinance to the target. The target employs kinetic and nonkinetic (NK) weapons to counter this threat. The use of real-time digital signal processing (DSP) algorithms greatly enhances the effectiveness of the modern threat sensor.

This work illustrates the present state of electronic warfare (EW), describing practical examples of DSP employed by the sensor of an autonomous threat for the purpose of countering standard electronic attack (EA) techniques. In addition, the means to assess EW performance by a combination of physics-based modeling and engineering intuition is demonstrated.

Throughout the book, examples from naval EW provide an understanding of the issues of the modern EW information battle. Since most works focus primarily on air EW, this book fills a gap in the literature by providing the basic fundamentals of naval EW while illustrating the more general, modern EW concepts. At the present time, the autonomous radar-guided ASM has a significant tactical advantage resulting from the much improved radar technology and the use of high-speed DSP capability.

This chapter provides background material for modern EW and introduces basic EW concepts. The first section briefly summarizes the history of naval EW. It illustrates the evolution of modern EA from exploiting technical flaws in the sensor to a true battle for information. The present ASM radar sensor can readily detect and measure guidance parameters from multiple targets. Simultaneously employing a variety of DSP algorithms, the sensor rapidly and reliably extracts measurement features for improved target classification in the sense of electronic protection (EP) of the sensor.

In the next section, several general conceptual EW battles are summarized. These simple battle scenarios will serve as basic examples of broader concepts throughout and are used to define standard EW terminology. The concept of probability of raid annihilation (PRA) is described to show how these simple scenarios serve as the fundamental pieces of a general analysis of a swarm attack including radar-guided ASMs against a naval battle group.

In the final section, several simplistic engagement examples are defined for later use in the book. Using these basic and simple examples the results of the mathematical analysis applicable to the more complicated engagement are more easily and intuitively understood.

1.1 Evolution of Naval EW

The naval fleet has been a means to project visible power from the sea into a geographical area controlled by enemy forces since ancient times. Often, sailing vessels could move personnel and equipment along waterways more easily than over land. An enemy, not in the immediate proximity of land fighting forces, could at times be attacked from the sea. This led to battles against naval forces from the land and between naval forces at sea. Combatants would battle from one ship to another by getting a vessel close enough to physically put fighting personnel onboard the opposing vessel.

In addition to armies, land weapons were often adapted to the sailing vessels to increase their war-fighting effectiveness or to improve the vessels' capability for self-defense. The ancient Greeks adapted various weapons to their vessels to make them effective over longer ranges. Examples of weapons used by the Greeks on sailing vessels include catapults, flame throwers, spears, and bows and arrows. With the introduction of gunpowder and cannons to breach castle walls and to throw ordinance to longer distances, these too were quickly adapted to sailing ships. With these technological advances, significant damage could be inflicted over ever greater distances.

During a battle, it is important to be able to communicate between different groups of combat forces. In ancient times, long-distance communications were limited to some audible signals and to line-of-sight visual signals.

Just as in land battles, naval vessels used flags and smoke to send information to other combatants. At night, signal fires were used.

This information battle is very important. In addition to coordinating tactics and forces, deceptive actions could turn a battle. Sometimes a vessel would not be known to have been captured by an enemy force. A false flag could be flown to deceive the enemy into mistaking the enemy for a friend, thus allowing the vessel to close the separation distance before being classified as an enemy. This is an early and simple example of deceptive information warfare.

Just as weapons and communications evolved and improved so did other technology. An important improvement for the naval warfare sensor is the use of the telescope to increase the distance at which a vessel could be visually inspected and classified. Improvements in navigation included the use of the compass and the sextant combined with improved knowledge of the stars. Improved navigation allowed the vessels to sail to greater distances and to venture farther from the shore.

With superior roads and a naval fleet, Rome was able to control the area around the Mediterranean Sea and beyond. With improvements in navigation, communications, sensors, and weapons technology, naval fleets could project power over ever greater distances. Great Britain was able to control a vast worldwide empire for many years by virtue of its superior navy. The United States used its fleet to impress the Japanese with a demonstration of advanced naval technology in 1853. This led to increased trade between the United States and the Asian nations.

As with these earlier technological improvements the invention of aircraft was adapted to land warfare and soon after was adapted to the naval fleet. Placing aircraft on a ship via the aircraft carrier enabled the fleet to project power over much greater ranges and deeper inland from the sea by carrying the aircraft along with a floating base and launching airborne weapons from this mobile base. As demonstrated at Pearl Harbor and other islands of the Pacific in World War II, the combination of air platforms with naval vessels became very effective.

The means to counter the naval fleet continued to be via land-based weapons, other ship weapons, and airborne weapons. In battles between naval fleets the ships had to get close enough to be within the range of onboard weapons. During World War II, this was at times accomplished by stealth via an attack by fleets of submarines. At other times, combatants using aircraft launched from aircraft carriers would attack the opposing fleets over considerable separation ranges. Attacking the fleet by aircraft meant having the pilot guide the air-to-surface ordinance to the fleet ships. Increased attacks from these pilot-guided weapons required ever-more sophisticated defensive weapons to be developed to provide the ships with adequate self-defense from airborne threats.

Also during World War II, radar matured as a sensor in land/air battle and then in sea battle. This electromagnetic-based sensor improved the ability to detect and locate (angle and range) enemy air forces and naval forces over much greater distances as well as through darkness and during inclement weather. Almost as soon as combatants began to use radar, the development and deployment of deception and confusion techniques began, such as the use of chaff. With the increasing reliance on radar, EW became a distinct aspect of combat.

As the speed, complexity, and range of battle increased along with the development of the radar sensor and other electronic technologies, greater automation of the battle evolved. For example, the operator could interpret the radar display screen for targets and chaff. Or electronic processors could highlight particular radar echoes to the operator as preferred potential targets of interest based on processing of the echo features. Eventually the display entirely removed the raw radar data and replaced it with digitally generated symbols. Increasingly the electronics-based weapon made tactical battle decisions or prompted the operator with real-time combat recommendations. Weapons were becoming more autonomous.

In the modern era, many nations have navies consisting of various ships that can project force over considerable distances along the seas adjacent to hostile nations. Several nations like the United States build fleets around aircraft carriers designated as the high-value unit (HVU). This fleet is customarily used to project power into a hostile region by the aircraft launched from this mobile base. For the United States, the aircraft carrier has become a dominant mobile weapon base both as a means to inflict damage and as a very visible means of making political statements. While the other ships can project weapons over long distances, their primary role is often to protect the HVU as the mobile base for the aircraft. And as the technology becomes more sophisticated, the cost of the fleet increases.

As the naval fleet is a means of projecting power into a geographic area controlled by an enemy force, and as combat is more automated, the natural counter to this naval power is the autonomous ASM. The ASM threat has developed for over 60 years as an alternative to pilot-guided weapons by combining various improved sensors with improved electronics and processors. The modern ASM can be launched by sea vessels, air platforms, submarines, and from land. While expensive, the ASM is a very cost-effective means of countering the ever-more expensive naval fleet; this is especially true if it can successfully target the HVU.

In 1967, Israel sought to project power to the Egyptian controlled Sinai Peninsula. Egypt launched four Russian built Styx missiles and sunk the Israeli navy ship INS Eilat. In response, Israel accelerated its development of the Gabriel class of ASM. These missiles were used effectively against Syria and others in the Yom Kippur War in 1973, sinking several enemy ships.

When faced with an enemy (Argentina) in the Falklands War in 1982, far from its main bases, Great Britain used its naval fleet to project its power many miles away from Britain into the ocean region near Argentina. In response, Argentina was able to effectively use the French built Exocet ASMs that were air launched and land launched. The Sheffield and two other British ships were sunk or damaged by these ASMs.

At this time, the U.S. Navy realized that its automated defensive systems would ignore the Exocet, classifying it as a nonthreat since it was a NATO ally weapon. The U.S. Navy defense system software was quickly modified to correct this when necessary. And during the Iran-Iraq conflict, the United States sought to project power into the Middle East against Iran by positioning naval vessels in the seas of the region. Although it was said to be pilot error, Iraq effectively used Exocet ASMs versus the USS Stark in 1987.

Even nations with limited forces can implement a reasonable counter to a superior navy through the acquisition and fielding of ASMs. During the Lebanon War in July 2006, Israel was containing Hezbollah with its navy patrolling about 10 nmi off the Lebanese coast. Not realizing Hezbollah had acquired ASMs (probably the C 802), the Israeli ship INS Hanit was not using automatic defenses lest they misidentified Israeli aircraft for threats. Reportedly, Hezbollah launched two ASMs at the ship from the coast, one of which disabled the ship and killed several sailors.

There are two basic aspects of fleet defenses. The first aspect is to protect the ships and their ability to project power. The second aspect is to protect the fleet in a hostile area for as long as possible. The primary means of protecting ships against ASM is via kinetic or hard-kill weapons. Examples of kinetic weapons include antimissile missiles such as Sea Sparrow and RAM, high-rate guns such as close-in weapons system (CIWS), and exotic newer weapons such as high-powered laser-beam weapons.

Kinetic weapons are the most reliable and effective defense against ASM. The goal of these weapons is to physically damage or destroy the ASM platform or its sensor. A virtue of kinetic weapons is the ability to confirm the kill in real time by sensing the ASM deviating from its path or, preferably, to observe the ASM crashing into the sea. A counter by the ASM to kinetic weapons is to maneuver during the attack. This greatly complicates the kinetic weapon fire-control algorithm.

Some nations with larger military forces such as the People's Republic of China or the Russian Federation are expected to field waves of ASMs to counter a hostile fleet. These can be launched from aircraft, land bases, surface ships, and submarines in a coordinated manner to overwhelm naval defenses. A wave attack of modern maneuvering ASM threats against a naval fleet presents a formidable challenge to fleet defenses. Such an attack is expected to overwhelm existing hard-kill assets.

Since the wave attack and the ASM maneuver are effective counters to kinetic weapons and the ability to carry these kinetic weapons is limited, it is sensible to augment kinetic weapons with NK weapons or soft-kill weapons. NK weapons can be used to counter multiple ASMs, and ships can generally carry more NK stores per volume of space. This necessitates an effective fleet EA or soft-kill defense capability as a complementary adjunct to hard-kill defense. Figure 1.1 simplistically illustrates the multilayered defense in naval battle against a wave attack of ASMs.

EW is an information battle [1]. The purpose of the ASM seeker is to gather and interpret information so that the ASM can make automatic tactical decisions. The first tactical task of the seeker is to detect all possible targets in the surveillance mode. The next task is to classify the target or to identify the correct target. Figure 1.2 shows a simplified model of the ASM platform.

After detecting and selecting a target the ASM sensor processing generates tracking gates around the target parameters to isolate this data and protect the sensor from inadvertently using false information from jamming systems. Once in the tracking or localization mode the seeker makes localization measurements (range and angle) of this target for the guidance subsystem. The goals of the ASM threat seeker are to make the correct target detection and

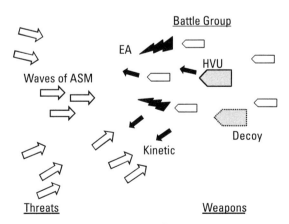

Figure 1.1 Multilayered naval battle against ASM threats.

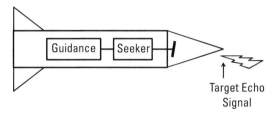

Figure 1.2 ASM platform model.

then to make accurate range and angle estimation while operating in the presence of jamming. Figure 1.3 illustrates the use of a simple range tracking gate.

While kinetic weapons attack the ASM platform, EA systems attack the ASM sensing functions by input of false information to the seeker. Figure 1.4 shows the ASM under attack from both kinetic and NK weapons.

The proper development of classical EA techniques required intelligence agencies to collect samples of the actual functioning threat sensor to discover a sensor hardware flaw. Examples of classical EA techniques that attacked the detection capability included the use of chaff as an alternative target or the use of noise jamming to blind the sensor and prevent or delay target detection.

For example, examining a functioning enemy ASM seeker enabled EW engineers to design an EA radar pulse sequence that would capture the tracking gates and then to position false data in these gates as illustrated in Figure 1.5.

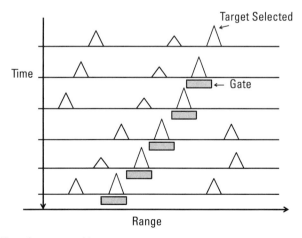

Figure 1.3 Use of range tracking gate.

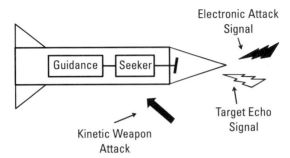

Figure 1.4 Attacking the ASM.

An example of damaging false target data was to generate a low-duty-cycle false target (countdown technique). This input would make the tracking loop unstable without the ASM system realizing it was interrogating a false target. Other examples of classical EA based on sensor technology flaws include AGC deception and blinking jamming. These techniques emphasize the need of the ASM to recognize the true target. Descriptions of these and other classical techniques can be found in many EW references [2]. These types of EA techniques are generally no longer effective.

Consider the autonomous ASM launched to sink a ship. The purpose of the ASM radar seeker is to detect the target as a radar echo and from this echo to measure the ship location (typically range and angle) for ASM guidance information. Range is estimated from the time delay of the echo. A guidance

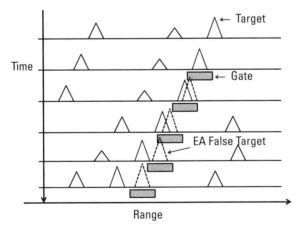

Figure 1.5 Range gate capture for seduction EA.

algorithm such as proportional navigation requires continuous localization measurements to properly control the ASM flight trajectory. Figure 1.6 illustrates the basic requirement of proportional navigation [3, 4].

The basic concept of proportional navigation is to keep the line of sight fixed from the ASM to the intended target. For example, consider the triangles formed by the trajectories and the look vector from the ASM to the target. It is seen from the diagram that if the triangles are all similar the ASM is on a collision course with the target. Once a target is detected the sensor must measure this line of sight or the corresponding line of sight rate. Any deviation of the instantaneous line of sight is used to derive a correction to the guidance system. Tracking gates are created to protect the sensor from inputting false data. Only data close to the originally tracked target is accepted. If the EA system can capture these tracking gates, false data can be input to the guidance loop.

Proportional navigation requires a measure of the rate of change of the pointing angle to the target. This rate of change of the pointing angle is related to the pointing angle error. The ASM design engineer needed to develop a sensor system that would measure the radar antenna pointing error. One of the early techniques developed to measure the pointing angle error (angle to the target relative to the antenna centerline) was the conical scanning (CON SCAN) tracking radar. The parabolic antenna beam is physically rotated about a central axis at a small offset angle relative to the antenna beam width. A target in the beam generates a time-modulated return with the modulation rate equal to the antenna rotation rate. The angle to the target relative to the rotation axis is estimated by measuring the phase and amplitude of the resulting modulation [5, 6]. As illustrated in Figure 1.7, the direction angle is related to the phase of the maximum of the modulated signal relative to the known antenna rotational position angle.

For adequate target detection, the ASM radar must transmit high-power signals. Since the scanning antenna causes the transmit beam to modulate this high-power pulse at the target, this antenna scanning is readily detected

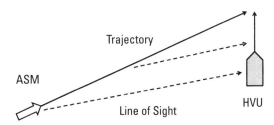

Figure 1.6 Proportional navigation.

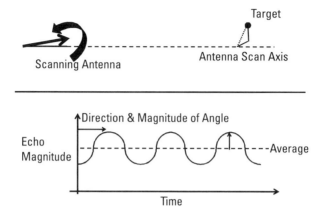

Figure 1.7 ASM seeker model of CON SCAN antenna.

and measured by the ship EW support system. This flaw in the sensor is then exploited to corrupt the angle estimated by transmitting a corresponding modulated jamming signal. Figure 1.8 demonstrates the impact of a 50% duty-cycle EA signal properly phased relative to the scanning threat antenna. Transmitting the EA when the modulated target signal is at a minimum at the ship support system, the direction angle is estimated to be in the opposite direction to the true target and at a large angle error.

With advances in radar technology, it became possible to electronically rotate the receive antenna while keeping the transmit signal fixed. This EP technique was implemented in the conical-scan-on-receive-only radar to counter the EA. Hardware imperfections still allowed a small receiver leakage

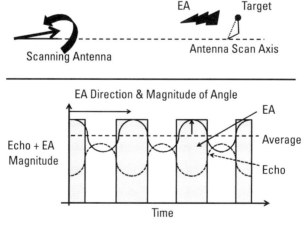

Figure 1.8 ASM CON SCAN antenna plus deception EA.

modulation to be imposed on the transmit beam. While more difficult to detect, this signal would again make possible an effective EA.

The final technological improvement in the angle estimation information battle was the advent of monopulse angle estimation techniques perfected in the 1970s [7]. Monopulse enables the estimation of angle pointing error with a single pulse and no time-modulation requirements. Rather than being spatially modulated, the transmit pulse is generated via a sum antenna pattern for maximum gain along the antenna bore sight. For example, the single feed of the parabolic antenna is replaced by a four-horn feed and a hybrid waveguide. A typical configuration is illustrated in Figure 1.9.

By transmitting energy through all four horns simultaneously, a constant beam power is transmitted. Using a series of magic T's, the receive energy can be combined as a sum beam and as various difference beams. The difference beams shown in Figure 1.9 correspond to an azimuth beam and an elevation beam. The difference beams provide information about the target location in angle relative to the antenna bore sight. Taking the ratio of a difference beam and the sum beam results in a direct measure of the angle off bore sight of the echo with a single pulse that also normalizes the measurement in range (this will be discussed in detail below). The central portions of typical beam patterns are illustrated in Figure 1.10.

In the original monopulse radar systems, the echoes received via both the sum pattern and the difference pattern were multiplexed into a single

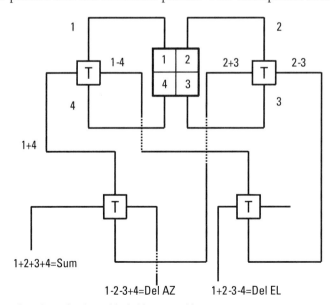

Figure 1.9 Four-horn feeds and hybrid waveguide.

receiver. The modern implementation of monopulse angle estimation is via the use of two coherent receivers. Angle deception is only possible through the corresponding use of two coherent EA antennas as illustrated in Figure 1.11. Some of the various monopulse EA implementations are designated as cross-eye, cross-pol, terrain bounce, and various combinations of these basic techniques, for example, double cross. Other than signals from this class of EA system, the angle to any single signal source can be accurately and reliably measured by the modern monopulse radar system.

As these examples illustrate, improvements in fleet-defensive systems are continuously countered by corresponding developments of new ASM capabilities. Since ASM radar sensors were originally X-band radar the EA systems' frequency coverage was restricted to X-band. This led the ASM engineers to develop and implement ASMs with Ku-band radar seekers. Since the primary threat was the sea-skimming ASM, the angle coverage of naval EA systems was restricted to low angles to reduce cost. This led the ASM engineer to

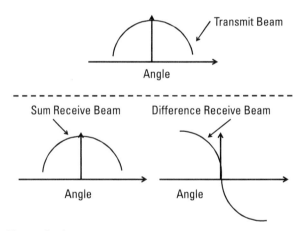

Figure 1.10 Monopulse beam patterns.

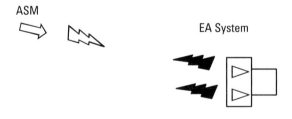

Figure 1.11 Dual coherent source EA.

employ ASMs with a pop-up maneuver. Next the ASM engineers developed the capability to fly a steep dive trajectory. The use of improved kinetic weapons to defend the fleet was countered by altering the guided course to include high-g maneuvers.

The modern ASM is fast and highly maneuverable. This platform can fly low and it can use sophisticated navigation subsystems to fly extended distances. It can fly low below fleet defenses and sensors and periodically climb to collect data over the horizon, update its guidance and targeting information, and then drop back down to continue its approach at sea skimming altitude. When exposed to kinetic weapons, the ASM performs high-g maneuvers to complicate the kinetic weapon fire-control algorithm.

The modern ASM seeker contains one or more sophisticated sensors and multiple high-speed digital signal processors. These technologically advanced sensors can be visual, infrared (IR), and microwave. Radar component technology has steadily improved, culminating in the low probability of intercept (LPI), monopulse radar sensor of today. The LPI radar transmits a very-low peak energy pulse that is wide in time and contains a coded waveform modulation. The radar is difficult for the battle group to detect, especially in its own complex radar environment. The modern LPI radar generates accurate digital data via analog-to-digital converters (ADCs) with 12 to 14 bits. This wide dynamic range plus an AGC circuit renders it very difficult to generate enough energy to saturate the sensor. The typical radar sensor uses two or more coherent radar receivers to detect the targets and measure angle and range with extreme sensitivity and accuracy. The LPI radar provides extremely sensitive target detection, with more accurate location capability, and greatly improved resistance to EA [1, 8].

The ASM radar sensor has evolved to the point of fundamentally altering the basic methods of EW. Whereas classical EA was primarily the exploitation of flaws in the sensor, modern EW is a true battle for information, both its exploitation and dominance [1]. The modern sensor for an autonomous threat performs the functions of target detection, localization, and classification. With greater sensitivity the ASM sensor can readily detect all of the targets in its view. The sensor detects all of the ships as well as any false targets (both passive decoys and actively generated decoys) within its view. The sensor can simultaneously and accurately extract localization and feature measurements from all of these targets. With extensive memory and sophisticated digital processors, all of this data can be examined in real time.

The modern radar will detect all the targets and can monitor the multiple targets simultaneously. It is the objective of the EA to convince the ASM sensor to derive guidance input data from the false target and to not track the ship target. The primary EA means of defeating the threat sensor with certainty is to present the threat sensor with a decoy that adequately mimics the features of the ship target. The modern EA is designed to either hide the ship target features and/or to enhance the off-board decoy target features. To simplify the view of the modern EW, the objective of the naval EW defender is to make the target look less like a target and/or to make a decoy look more like a target.

Since all of the targets can be readily detected and located, the probability is high that the ASM will hit the chosen and tracked target. The only issue for the EW battle is the probability of the ASM selecting the correct target, whether the HVU or one of the other ships in the fleet. To select the correct target, algorithms in the digital processors must sort through the identifying features of the multiple ship targets and EA-generated false targets. This sorting function is performed in the presence of various EA, including the generation of false targets. To mitigate the detrimental effects of EA on the information, the sensor operation includes countermeasure techniques designated EP.

Thus, EA can attempt to deny target inputs or to provide false deceptive inputs. Deception techniques include providing false targets either electronically or via an actual false target. Examples of false targets include chaff, floating or airborne reflectors (passive decoys), and airborne or floating electronic false echo transponders (active decoys) such as Nulka. Another means of deception is to attempt to blind the sensor by generating high levels of electromagnetic noise. If the false target or noise is generated from a ship, then the EA system must eventually seduce the ASM sensor tracking to an off-board false target. If this is not done the ASM sensor will eventually detect and target the ship.

The modern EW battle task of primary importance is now target classification. Since EW is primarily classification deception via off-board false targets, an example of EP employed by the ASM seeker is to implement coordinated multiple sensors. This forces the EA defender to generate coordinated false targets to multiple and distinct ASM sensors. For example, suppose the ASM active seeker radar detects and locates a target echo and a decoy echo. At the same time, the ASM seeker passive radar detects onboard ship radar emissions at the same bearing as one of the targets as shown in Figure 1.12.

The modern sophisticated ASM radar sensor with multiple coherent receivers extracts target classification features in the presence of noise jamming and multiple false targets from the fleet. This digital data is then analyzed by highly sophisticated DSP via high-speed processors. Probabilistic mathematical techniques are used to determine the optimal target. In this way, the ASM sensor has evolved to achieve a significant tactical advantage.

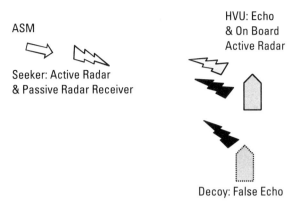

Figure 1.12 ASM use of coordinated sensors.

The ASM EP techniques protect the ASM from being misled by false targets. The main purpose of this work is to describe this capability of the modern ASM radar sensor to classify targets. The goal is to give the reader an intuitive understanding of the basic EP technical issues. This work explains DSP algorithms that are most useful for an autonomous sensor to quickly and accurately perform target classification decisions during its engagement with EA weapons. Throughout the book, a simple physics-based mathematical model is developed and described to guide the reader's understanding.

Understanding the material contained herein is essential to focus the efforts of EA development engineers to make the ship target appear less like a ship target and/or to make a decoy appear more like a ship target. In addition, understanding this material is essential to guide intelligence personnel on where to focus their information gathering efforts to accurately assess the ASM sensor EP capabilities.

This change in emphasis applies to all aspects of modern EW, but is particularly evident in naval EW. For these reasons, and because most works of EW focus on airborne EW, particular points throughout this book shall be demonstrated by the example of a radar-guided autonomous ASM attacking a ship. This is especially important at this time because the autonomous ASM presently enjoys a significant tactical advantage [9].

1.2 Terminology and Model Scenarios

The scenario studied in this work is a naval fleet of ships or a battle group projecting force into a geographical area controlled by opposing combatants.

These combatants attack the fleet with waves of autonomous ASMs. A typical fleet consists of at least one aircraft carrier together with an assortment of escort ships. The ships are termed the targets. The primary target is the aircraft carrier designated the HVU, while the escort ships possess the capability to launch ordinance to a considerable distance; their primary role is to protect the carrier or HVU as a mobile base for aircraft.

Various nations have developed and continuously improved the ASM for more than 60 years. The ASM is designated the threat. It can be launched from land, air, sea surface platform, and subsurface platform. The ASM may be launched from close to the fleet or from a considerable distance. With modern navigation systems, the ASM can fly several waypoints en route to the fleet if necessary. It is expected that the attack on the fleet will occur in waves of multiple ASMs over an extended period of time and from several directions. These ASMs will use a variety of flight profiles and sensor configurations. The high-diving ASM has the added advantage of only accepting targets on the sea surface. That is, a target evaluated as being above the sea surface or below the sea surface is obviously a false target.

The sea-skimming ASM is the primary example of threat discussed herein. This simplifies the task to basically a two-dimensional problem. This simplification does not significantly alter the target classification task of the ASM. At about 20 km, the sea-skimming ASM typically pops up above the sensor horizon for a quick target reacquisition and then conducts the terminal phase of its attack.

The ASM seeker contains one or more sensors designed to gather information used by the ASM guidance subsystem to guide the ASM to the selected target ship. The purpose of the seeker is to detect and classify the desired target and to estimate location information for the ASM guidance system. The ASM sensors include radar, IR, electro-optical, and various combinations of sensors (multisensor seekers). The ASM sensor suite also includes an altimeter. In this text, the ASM seeker sensor of interest is a coherent pulse Doppler radar.

With improved sensors the ASM can seek out the ship target and autonomously provide accurate guidance measurements. An individual ASM may seek to attack the HVU, an escort ship, or any ship target of opportunity. In this way the goal is to diminish the battle group effectiveness and/or the overall defensive capability of the fleet.

The defense of the ships is an orchestrated combination of weapons, including hard-kill or kinetic weapons and soft-kill or NK weapons. Examples of kinetic weapons are various antimissile missiles such as RAM and guns such as CIWS. These weapons attack the body integrity of a threat. The goal of a kinetic weapon is to physically damage the ASM and result in failed trajectory or sensing.

The emphasis in this section is on NK weapons or EW. NK weapons attack the seeker sensors. This is EA. The classical radar ASM seeker of the past is now replaced with radio frequency (RF) hardware that collects low-power coherent data via multiple receiver channels for rapid and effective DSP. With present radar technologies the modern ASM threat sensor can readily detect and accurately locate multiple targets. These potential targets are continually subjected to advanced DSP techniques for target feature analysis.

This information gathering capability of the ASM modern radar sensor coupled with high-speed digital signal processors leads to a significant shift in emphasis of EW. The emphasis for the threat radar engineer is now on EP in the sense of target identification and classification via target feature extraction and analysis. With the implementation of these practical digital EP techniques the modern radar sensor can quickly and reliably identify the correct target while collecting very precise guidance measurements. This improved sensor capability renders current EA techniques obsolete and affords autonomous threats an unprecedented tactical advantage. This change in emphasis applies to all of modern EW, but is particularly evident in naval EW. The terminology for this book is summarized in Table 1.1.

Table 1.1
Terminology

Terms	Clarifying Terms
Targets	Ships
Primary target	HVU (Aircraft carrier)
Weapons	EA assets
	On-board EA
	Off-board EA (e.g., decoys)
	Kinetic assets (e.g., CIWS and RAM)
Threats	ASMs
Seeker sensor	Coherent LPI radar

NK weapons attack the threat sensor by adding signals to the HVU echo and other ship echoes in an attempt to corrupt and/or mask the analysis of the true signal. The goal of the EA is to input information to the ASM sensor to make selecting a ship as a target less likely and to make selecting a false target more likely. The ultimate goal is to make the ASM sensors select an off-board decoy as the target. The measures taken by the threat sensor to counter this EA are EP. As stated above, EP is the main subject of this work.

The performance capability of kinetic weapons to counter an ASM is estimated based on a combination of intelligence, modeling, and testing. This is the hard kill a priori probability of kill. This estimate plus factors such as the number of assets available, mission goals, and expected battle progression are used to make an initial decision by the fleet warfare management system as to which weapon to use and when to fire. The result of using the weapon is observed for information about its real-time effectiveness. The observation can be via radar, IR sensor, or visual observation. This estimate of effectiveness is the a posteriori or real-time probability of hard kill. If the weapon is deemed successful, no other weapons are deployed. If it is assessed that the weapon was not successful, additional weapons can be used in a continued and layered attempt to counter the ASM.

Since a wave attack of ASMs can overwhelm kinetic weapons and the ASM seeker can counter most classical EA, fleet-defensive EW must evolve into a reliable and viable weapon system as an adjunct to the kinetic weapon system. The capability of NK weapons to counter an ASM must also be estimated based on a combination of intelligence, modeling, and testing. The weapon fire control system requires this NK weapon a priori probability of kill. This estimate plus factors such as the number of assets available, mission goals, and expected battle progression must be used to make an initial decision by the fleet warfare management system as to which weapon to use and when to fire. The result of using the weapon must be observed for information about its real-time effectiveness. This estimate of effectiveness is the a posteriori or real-time probability of soft kill. If the weapon is deemed successful, no other weapons need to be deployed. If it appears the weapon was not successful, additional weapons can be used in a continued and layered attempt to counter the ASM.

To develop this defensive capability the EW engineer must understand the ASM digital processing techniques. The main purpose of this work is to describe the common techniques used by the modern ASM radar sensor to classify targets. The goal is to give the reader an intuitive understanding of the basic EP technical issues. This text explains the current DSP algorithms that are most useful for an autonomous sensor to quickly and accurately make target classification decisions during its engagement while deliberately countering standard EA weapons. Throughout the book, a simple physics-based mathematical model is developed and described to guide the reader's understanding.

The soft-kill weapons are various radar EA systems. The EA system may be onboard the HVU or off-board the HVU. If the jamming system is off-board the HVU, it may be on an escort ship or on another platform, for example, a drone or a decoy. The jamming system can be active or passive. An active jamming system electronically generates a radar signal. The active EA can modify the ASM radar waveform signal it receives for retransmission back to the ASM radar. The jamming system can generate the attributes of false target signals or some other EA signals, for example, noise jamming. A passive jamming system uses some type of reflector to generate an echo of the ASM radar sensor transmitted signal. Examples of passive jamming include corner reflectors and chaff [1, 2, 5].

The ASM can be represented as consisting of several basic subsystems. The propulsion system and aerodynamic body make the vehicle able to fly and maneuver as required. The guidance system generates control surface commands and propulsion system commands. The examples in this work will generally be for a sea-skimming (low flying) ASM.

The seeker contains one or more sensors. The seeker goal is to detect and classify the proper target. The seeker then measures localization information from the target of interest for input to the guidance system. The sensor for examples in this work is the coherent LPI radar with two channels or monopulse receivers and multiple digital signal processors. The main seeker components are summarized in Table 1.2. The seeker antenna sensor must be enclosed in an aerodynamic radome. The overall processing control of the ASM is with a general digital processor.

The threat (ASM) has a prescribed timeline or sequence of events that it plans to execute as it prosecutes the target (HVU) for the purpose of delivering its ordinance to the HVU. Figure 1.13 illustrates this sequence.

Table 1.2
ASM Seeker Model Primary Sensor Components

Coherent LPI radar
Flat plate monopulse antenna in an aerodynamic radome
Multiple receivers
ADCs
Multiple DSPs

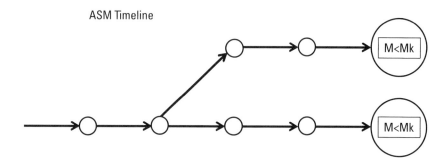

—M is the ASM to ship miss distance
—Mk is the design goal

Figure 1.13 ASM preferred timeline.

At one or more points on this timeline, a weapon (EA) can be used in some way (defined as an action) to alter the threat timeline in a manner desired by the fleet defender. The EA goal is to have the threat follow a sequence of events that mitigate its performance and cause it to not hit the target ship. This timeline sequence is defined as the weapon strategy. This is illustrated in Figure 1.14.

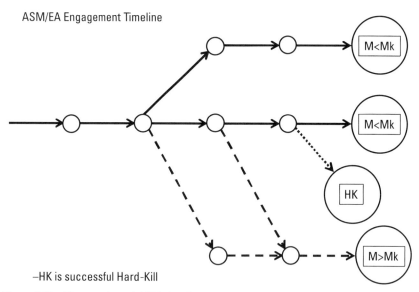

—HK is successful Hard-Kill

Figure 1.14 Fleet preferred ASM timelines.

Associated with this preferred sequence is the a priori probability of defeating the threat. This probability is the EA probability of kill. This probability must be estimated from extensive modeling, simulations, and testing based on various forms of intelligence. A simple architecture of the fleet-defensive system is shown in Figure 1.15. This topic will be discussed more completely in Chapter 7.

Thus, there are multiple timelines of events and actions, and the resulting timeline determines whether the actual threat hits the ship or not. At each decision point on the timeline, there are many observables available from the many sensors available to the fleet. These observables are evaluated to determine the a posteriori probability of defeating the threat or the real-time probability of EA effectiveness. This is a real-time estimate of the probability of kill for a particular engagement.

There are two basic types of real-time assessment that may be required during an engagement. One type of assessment deals with evaluating the state of the ASM over an extended period of time. For example, evaluating the non-maneuvering ASM trajectory leads to an estimate of the miss distance. Another type of assessment is to evaluate a change of state at the time of an action. For example, if the ASM is tracking the ship, it will generally point its antenna at the ship. If an action by a weapon is executed for the purpose of making the ASM change to tracking a decoy one may examine observables to sense a change in the aim point of the ASM antenna such as antenna jog detection.

In the next section, it will be shown that the defense analysis can be understood through the analysis of a single ASM against a single ship. The capability to defeat the threats through combinations of weapons is defined by the PRA for a particular strategy.

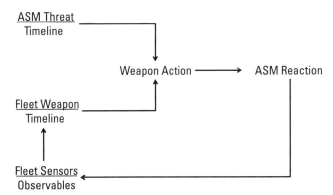

Figure 1.15 Fleet-defensive system architecture.

1.3 Probability of Raid Annihilation

To optimize the probability of fleet survival and the time on station, the U.S. Navy has developed fire-control algorithms based on the PRA. The PRA is defined as the ability of the stand-alone ship as an integrated system with the battle group to detect, control, engage, and defeat a specified raid of ASM with a specified level of probability in an operational environment. The ideal goal is a PRA of 1.0, definite counter of the ASM raid. For simplicity, assuming a single-threat type, assuming nK shots of a single kinetic weapon type (K), and assuming nNK attempts with a single NK weapon type, the PRA for a raid of the number (NT) of identical threats is

$$\text{PRA} = \left\{ P(D) \cdot P(E|D) \cdot \left[1 - \left\{ 1 - P(K|E) \right\}^{nK} \cdot \left\{ 1 - P(NK|E) \right\}^{nNK} \right] \right\}^{NT} \quad (1.1)$$

In this expression, $P(D)$ is the probability of detecting the threat and $P(E|D)$ is the corresponding probability to engage the detected threat. The other probabilities are the probability of weapon success in defeating the threat given the decision to engage. The PRA is a probability between zero and one. If the threats are detected, engaged, and defeated (all Ps equal to 1), then PRA is equal to 1 and no ASM threats are successful in damaging any ships. In this book, the focus is on the last probability: $P(NK|E)$.

Any number within the brackets less than one reduces the PRA to less than 1. Then, as NT increases, the PRA decreases. The PRA (probability to counter the threats) decreases as more threats are added to the attack. Obviously, an inability to detect the threats decreases the probability of ship survival. In addition, if a threat is detected but not engaged, the PRA decreases. If the ASM uses LPI radar sensor, it is difficult to detect its radar emissions. As the ASM approaches fast and low, it is difficult to detect the ASM platform until the terminal phase. Assume that the ASM is detected and engaged. That is, the detection probability is equal to 1 and the engagement decision probability is equal to 1. With these simplifications

$$\text{PRA} = \left\{ 1 - \left\{ 1 - P(K|E) \right\}^{nK} \cdot \left\{ 1 - P(NK|E) \right\}^{nNK} \right\}^{NT} \quad (1.2)$$

The goal is for the PRA to be as close to 1 as possible. If a defensive weapon is employed and it is absolutely effective in defeating the ASM, then one of the probabilities on the right hand side is equal to 1. If either probability of kill for an individual employed weapon is 1, then PRA is 1, as desired. Consider the use of one weapon type, say the kinetic weapon.

$$\text{PRA} = \left\{1 - \left\{1 - P(K|E)\right\}^{nK}\right\}^{NT} \qquad (1.3)$$

If the a priori probability of kill for the weapon [$P(K|E)$] is 0.7 and a single shot (nK = 1) at a single threat (NT = 1) is executed, then PRA = 0.7. However, if there are four shots (nK = 4) at the single threat, then PRA = 0.9919. Taking more shots at the threat improves the probability of survival, but depletes resources needed for other threats and for the defense against the continued wave attack.

Assume that there are four threats in the wave (NT = 4). If there is a single shot at each threat, this would expend four weapons and the PRA would be 0.24. Again taking more shots can improve the PRA. Suppose there are four shots at each of the four threats. This expends sixteen weapons and the PRA is 0.97.

Once the defensive weapons are depleted, the fleet must resupply weapons stores or retreat. This translates to less time on station since these weapons must be replenished to be able to protect the fleet ships.

Algorithms have been developed by the U.S. Navy to automatically select combinations of weapons and tactics to optimize both survivability and time on station. These algorithms require a priori probability of kill for both kinetic and NK weapons. In addition, survivability is improved by accurate estimates of time varying a posteriori probabilities of kill for both weapon types. This latter probability relies on real-time estimation of parameters relating to the effectiveness of defensive actions. Mixing kinetic and NK weapons improves the PRA, but depletes resources and reduces the time on station.

Assume that one shot at each of four threats, for this example, of a probability of kill equals to 0.7. As stated above, the PRA equals 0.24. If real-time observations show that three ASMs are defeated, the a posteriori probability of kill for these three each is 1. Assume real-time observations show that the last threat is not defeated. To defeat the last threat, ASM additional weapons are deployed. If two additional weapons are deployed, the PRA is now 0.91. Again measurements must be made to refine this estimate in real time until the PRA is equal to 1.

For the purpose of this book, the primary interest is the PRA contribution from NK weapons. Repeating (1.2)

$$\text{PRA} = \left\{1 - \left\{1 - P(K|E)\right\}^{nK} \cdot \left\{1 - P(NK|E)\right\}^{nNK}\right\}^{NT} \qquad (1.4)$$

Survivability depends on detecting and engaging the threats. Survivability is improved by a judicious combination of multiple engagements with both kinetic and NK weapons. Consider the example above of a single threat engaged with a single kinetic weapon with a probability of kill of 0.7 leading to a PRA of 0.7. If EA is used instead with a priori probability of kill of 0.5, then the PRA is 0.5. However, if both the kinetic weapon and the EA are employed (and assuming the use of the two weapons is independent), the PRA is 0.85, better than the use of either weapon alone as illustrated in Figure 1.16.

The focus in this text is on the contributions of the NK weapons. Thus, the primary interest is $P(NK|E)$. Consider the expression for a single threat (NT = 1) and NK weapons only

$$\text{PRA} = 1 - \left\{1 - P(\text{NK}|E)\right\}^{n\text{NK}} \qquad (1.5)$$

In addition, assume a single weapon engagement (nNK = 1). Then

$$\text{PRA} = P(\text{NK}|E) \qquad (1.6)$$

For a single shot of a single NK weapon, the two probabilities are equal. An expression for this probability is needed to examine this probability more closely.

In the IDECM program, an expression was defined for the reduction of threat lethality. For PRA analysis, the NK probability of success is defined in a similar manner. Thus, $[P(NK|E)]$ is defined

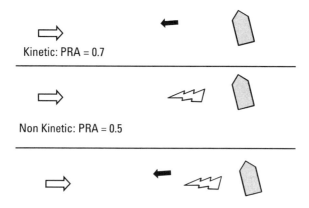

Figure 1.16 Simple PRA examples.

$$P(\text{NK}|E) = 1 - \left[\frac{P(k|\text{EA})}{P(k|0)}\right] \quad (1.7)$$

where $P(k|\text{EA})$ is the probability of ASM defeating the ship in the presence of EA and $P(k|0)$ is the probability of ASM defeating the ship when there is no EA present. From the definition of PRA, it is seen that this is an expression defining the lethality reduction of the ASM attack. For example, assume that the EA has no effect on the ASM. The two probabilities on the right-hand side are then equal and PRA = 0. Note that this does not mean the ship sinks. A PRA = 0 means that the threat ASM raid is not diminished. Or the EA employed has no impact on the ASM mission performance.

As another simple example, assume the area consists of a single ship and a perfectly effective decoy in the sense of mimicking the ship. Assume also that the ASM definitely targets either the ship or the decoy. Let P_d be the probability that the ASM targets the decoy and P_s be the probability that the ASM targets the ship. With these assumptions

$$1 = Ps + Pd \quad (1.8)$$

The probability that the ASM defeats the ship when there is no defense is

$$\text{Probability ASM kills ship} = P(k|0) \quad (1.9)$$

When the decoy is deployed, the probability that the ASM kills the ship is the probability that the ASM targets the ship times the probability that the ASM kills the ship. From the simple assumptions above

$$P(k|\text{EA}) = (1 - Pd) \cdot P(k|0) \quad (1.10)$$

For this example, the PRA is equal to the effectiveness of the decoy or the probability that the ASM will target the decoy. Again the PRA is the probability that the ASM effectiveness is reduced. Inserting this expression into (1.7) above for this example

$$\text{PRA} = P(\text{NK}|E) = Pd \quad (1.11)$$

As a final example, assume that the a priori (or deductive) probability is P_d = 0.95. Upon detection of the ASM, the defensive engagement is begun and the decoy is deployed and PRA = 0.95. As the ASM approaches the region (see Figure 1.17), suppose that its trajectory is measured and found to be consistent with the ASM targeting the ship with a probability 0.99. At this

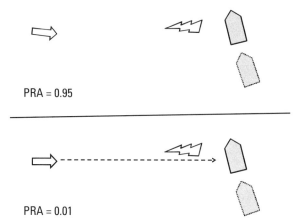

Figure 1.17 Simple PRA example: Decoy only.

point, the a posteriori (or inductive) probability is P_d = 0.01 and the PRA is now estimated to be PRA = 0.01, as shown in the lower panel of Figure 1.17. It would be prudent to initiate another defensive action such as the use of a kinetic weapon. Thus, it is seen necessary to continually update the real-time probabilities of weapon effectiveness during the engagement. More will be discussed on this subject in Chapter 7.

1.4 Sample Strategies

Throughout this text, it will be assumed that the ASM swarm attack includes the sea skimming, LPI radar-guided autonomous ASM. The focus is on the EW portion of the battle. The ship defense consists of varieties of self-defense and escort defense. Self-defense is defense of the platform with onboard weapons. These weapons are typically repeater EA systems. Escort defense is the defense of the ship with weapons onboard another ship or with an off-board decoy. The EA from an escort ship is typically an electronic repeater EA. At the present time, the off-board EA systems are autonomous. The off-board defense can be a passive decoy or an active decoy (repeater). Passive decoys may include combinations of corner reflectors or chaff.

The ASM has a preferred timeline sequence. The use of defensive weapons refers to the strategy of attacking the ASM through one or more EA actions or tactics. A sequence of these tactics shall be defined as a strategy. The goal of the action is to shift the ASM timeline to a sequence preferred by the ship. The ultimate goal of the EA is to have the ASM seeker guide to an off-board false target or, as a minimum, for the ASM to not track any target. The off-board false target is the ultimate means of defeating the ASM. It renders the ASM to guide to a false target, that is, to miss the true target. Figure 1.15 illustrates the architecture.

The subgoal of each EA action is to either make the target less likely to be tracked or to make an off-board target more likely to be tracked. As elaborated in the following chapters, there is a mix of potential means of defense. The onboard EA can either project one or more false targets or project cover noise jamming. The intent of the false target is to present an alternative target to the ASM sensor. The intent of cover noise EA is to mask the true target features. The goal of false targets from onboard weapons is to capture the tracking gates as a part of the strategy to seduce the ASM into tracking an off-board target.

For this work, three basic defensive EA actions of ship defense against an ASM are considered as simplistic examples to illustrate the general principles. Variations of these are discussed throughout as appropriate.

Strategy 1 consists of the HVU and a single decoy. This will be the most basic strategy. The decoy may be a combination of passive reflectors or an active repeater, or chaff. The goal of Strategy 1 is to capture the ASM tracking gates on the false off-board target from the onset of the ASM seeker doing detection, classification, and localization. Strategy 1 is illustrated in Figure 1.18.

Strategy 2 consists of the HVU and/or an off-board platform using some form of active EA to defend the HVU. Strategy 2 includes the possibility that the ASM seeker is tracking a ship target. In this case, one or more electronically generated false targets are used in an attempt to capture the tracking gates and to seduce the tracking gates from the ship and ultimately onto an off-board false target (Strategy 1). The generation of multiple false targets may be an attempt to add confusion to the scenario. While the EA attacking the ASM sensor (onboard and/or off-board EA) will include some form of

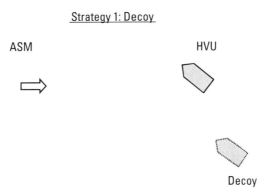

Figure 1.18 Defensive Strategy 1—simple use of a decoy.

electronically generated false targets, the scenario may include cover noise in the sensor field (Strategy 3). For example, this strategy may include the use of electronic chaff projected from a ship (the HVU or another ship) to the HVU location. In this case the ASM sensor would tend to reject the HVU since it displays characteristics of chaff rather than of the HVU. Strategy 2 is illustrated in Figure 1.19.

Strategy 3 is to cover or hide the true targets in an effort to deny the ASM seeker sensor from collecting any useful target data. If an off-board device generates cover noise jamming, it can seduce the ASM by concealing all targets and putting the ASM in home on jam or angle track only. If the cover jamming is generated by a ship, then it must be combined with some additional means to seduce the ASM to switch to an off-board device later in the engagement. The full scenario may include onboard EA consisting of cover jamming and/or some combination of electronically generated false targets as well as a decoy. Figure 1.20 illustrates this strategy.

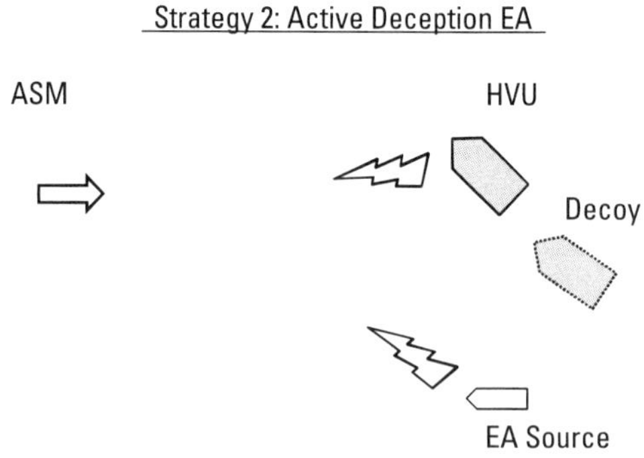

Strategy 2: Active Deception EA
ASM
HVU
Decoy
EA Source

Figure 1.19 Defensive Strategy 2—active EA protection.

Strategy 3: Cover Jamming EA
ASM
HVU
EA Source
Decoy

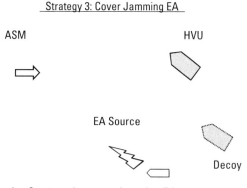

Figure 1.20 Defensive Strategy 3—cover jamming EA.

References

[1] Schleher, D. C., *Electronic Warfare in the Information Age*, Norwood, MA: Artech House, 1999.

[2] Van Brunt, L. B., *Applied ECM*, Dunn Loring, VA: EW Engineering, Inc., 1978.

[3] Shneydor, N. A., *Missile Guidance and Pursuit*, Chichester, U.K.: Horwood, 1998.

[4] James, D. A., *Radar Homing Guidance for Tactical Missiles*, Hong Kong: MacMillan Education, LTD., 1986.

[5] Stimson, G. W., *Introduction to Airborne Radar*, El Segundo, CA: Hughes Aircraft Company, 1983.

[6] Wiley, R. G., *Electronic Intelligence: The Analysis of Radar Signals*, Norwood, MA: Artech House, 1993.

[7] Sherman, S. M., and D. K. Barton, *Monopulse Principles and Techniques*, 2nd ed., Norwood, MA: Artech House, 2011.

[8] Pace, P. E., *Detecting and Classifying Low Probability of Intercept Radar*, Norwood, MA: Artech House, 2009.

[9] Stavridis, J., *Sea Power*, New York, NY: Penquin Press, 2017.

2

Pulsed Doppler Radar Basics

The purpose of the radar seeker of the autonomous ASM is to provide accurate target localization measurements to the ASM guidance system. This is the seeker-tracking function. A viable guidance solution enables the ASM to reliably deliver its ordinance to the chosen target. Given the various levels of prior information, the seeker sensor must first detect and isolate the target radar echo. This is the search function. The modern pulsed Doppler radar coupled with fast digital signal processors can reliably detect and track multiple targets. These potential targets include the high-value unit, multiple other ships, various active and passive false targets, and other jamming signals such as cover noise. Thus, the critical function of the seeker is to classify the targets and to identify the correct target. This is the classification function. To perform the classification function, various features of the echo signals from potential targets are measured and analyzed via multiple high-speed digital processors. Over the years, a variety of practical algorithms have been developed to reliably and rapidly identify the correct target in the presence of jamming or EA.

To better understand these rapid and practical classification algorithms the basics of the radar sensor must be understood. The simplest description of the ASM pulsed Doppler radar sensor is that the radar transmits a sequence of pulses of electromagnetic radiation with known characteristics. Each pulse reflects from the various objects. These reflected pulses, mixed with jamming system energy, are received by the radar. This received energy plus noise is processed by the sensor and converted to digital measurements. Information extracted from these measurements is used to identify the desired target and to estimate particular parameters for the purpose of guiding the ASM to impact the target. Since the fleet attempts to deliberately mask the target from the ASM seeker sensor and/or to deceive the ASM seeker sensor in other ways, EP techniques are implemented in the sensor to protect it from these EA actions. Beyond the standard description of the radar processing, the true worth of the modern ASM sensor is in the EP techniques made possible by practical and rapid DSP. These EP techniques are the primary subject of this book.

Before delving into a description of these modern EP techniques, the engineer needs a basic understanding of the ASM radar sensor. Fortunately, there are many excellent works describing pulsed Doppler radar. In this book, some of the basic characteristics of the pulsed Doppler radar are summarized for later use.

In the first section of this chapter, a simple RF pulse is described. To understand a pulsed signal receiver, the concepts of matched filter, carrier frequency, and complex phase are introduced. The concept of coherent radar signals is described. It is important to realize that the modern ASM radar sensor uses coherent technologies and practical DSP techniques.

In Section 2.2, a quick review is presented of DSP gain. The distinction is detailed between coherent processing gain and noncoherent processing gain. As an important example of this distinction, the concept is described of radar pulse compression. This aids the engineer in understanding the value of generating a very wide (in time and frequency) pulse characteristic of the modern low probability of intercept (LPI) radar. Simple mathematical analysis shows the connection between antenna beamforming concepts and range and other matched filter concepts.

The receiver dynamic range and the analog-to-digital converter (ADC) are described in Section 2.3. The ADC is the link between the analog radar processing and the DSP. The high number of bits of the modern ADC makes it very difficult to saturate the ASM sensor processor.

The remaining sections introduce several additional features of the sensor, including the antenna and its polarization as well as the Doppler processing. Both the antenna and the Doppler processing are examples of coherent gain. The concept of monopulse angle measurement is detailed. With the basics of this chapter, a simple mathematical model representation of the ASM digital radar data is developed in Chapter 3. This model is then used to elucidate the various classes of EP algorithms. In this manner, the engineer can gain an intuitive sense of the means of the ASM seeker to negate classical EA techniques.

2.1 Electromagnetic Pulse

Maxwell showed many years ago that electromagnetic radiation consists of an oscillating electric field orthogonal to an oscillating magnetic field that propagates at the speed of light [1, 2]. Both fields are transverse to the direction of motion of the energy. As an example, Figure 2.1 shows an oscillating electric field in the x-direction and an oscillating magnetic field in the y-direction. The radiation propagates in the z-direction at the speed of light. Since the fields are transverse to the direction of propagation, the electromagnetic propagation is represented as a transverse wave.

This time representation of the oscillating electric field at the spatial origin can be represented as

$$\overrightarrow{e(t)} = E_0 \hat{x} \cos(2\pi f_0 t) \qquad (2.1)$$

In this expression, the field has amplitude E_0 in the x-direction and oscillates at frequency f_0. This wave is said to be linearly polarized in the x-direction or vertically polarized as drawn. The electric field for radiation in the z-direction may have independent components in both the x- and y-directions. For radiation with electric field of magnitude E_0, a more general representation for the electric field solution to Maxwell's equations can be written as

$$\overrightarrow{e(t)} = E_0 \{\hat{x} \cdot a \cdot \cos(2\pi f_0 t) + \hat{y} \cdot b \cdot \cos[2\pi(f_0 t + \varphi)]\} \qquad (2.2)$$

$$1 = a^2 + b^2 \qquad (2.3)$$

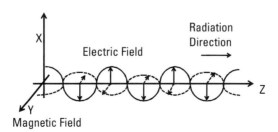

Figure 2.1 Simple electromagnetic radiation.

The parameters a and b define the magnitude of the x and y components, while φ defines the relative phase between the components. The polarization of the radiation is defined by viewing the electric field vector from the positive z-axis (looking at the approaching electromagnetic field). The general polarization is elliptical. Particular special polarization examples are illustrated in Figure 2.2. The polarization may be represented as a two-dimensional vector. A more detailed description of the polarization can be found in [3].

Neglecting the polarization, the time dependence at the spatial origin of the electric field amplitude in (2.1) can be written in complex form using Euler identities.

$$e(t) = \frac{E_0}{2}\left[e^{2\pi i f_0 t} + e^{-2\pi i f_0 t}\right] \tag{2.4}$$

The frequency can be understood as characterizing the spectrum of the signal. The spectrum of the signal represents the signal in frequency space via the continuous Fourier transform [4]. Given a time-varying function $s(t)$, its frequency representation $S(f)$ is

$$S(f) = \int e^{-2\pi i f t} s(t) dt \tag{2.5}$$

The inverse Fourier transform for well-behaved functions is

$$s(t) = \int e^{2\pi i f t} S(f) df \tag{2.6}$$

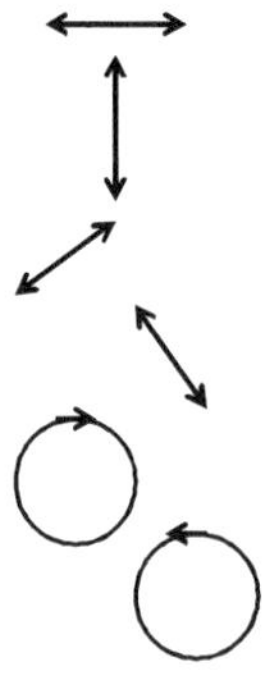

Electric Field	a	b	φ	Polarization
↔	0	1	—	Linear: Horizontal
↕	1	0	—	Linear: Vertical
↗	$\sqrt{2}/2$	$\sqrt{2}/2$	1/2	Slant Right
↙	$\sqrt{2}/2$	$\sqrt{2}/2$	0	Slant Left
◯	$\sqrt{2}/2$	$\sqrt{2}/2$	1/4	Left Circular
◯	$\sqrt{2}/2$	$\sqrt{2}/2$	−1/4	Right Circular

Figure 2.2 Special examples of polarization.

This can be verified (assuming certain smoothness properties of the functions) by substituting (2.5) into the expression

$$s(t) = \int e^{2\pi i ft} \left[\int e^{-2\pi i ft'} s(t') dt' \right] df \qquad (2.7)$$

$$s(t) = \left\{ \int s(t') dt' \left[\int e^{2\pi i f(t-t')} df \right] \right\} = s(t) \qquad (2.8)$$

The expression in the last brackets is the Dirac Delta function $\delta[2\pi f(t - t')]$ or the impulse function. It is zero for all values except at 0 where it is infinity. The integral of the delta function with any other functions equals the value of that function at 0.

Using this representation for the simple electric field in (2.4) and using the Dirac Delta function, the spectrum of this signal contains energy at both positive and negative frequencies as shown in Figure 2.3.

$$E(f) = \frac{E_0}{2} \delta(f - f_0) + \frac{E_0}{2} \delta(f + f_0) \qquad (2.9)$$

The above expression represents the electric field of the radiation at the source assumed to be at $z = 0$. At a distance R from the source along the z-direction, the electric field is

$$e(t, R) = E_0 \cos\left([2\pi f_0] t - \left[\frac{2\pi}{\lambda_0}\right] R \right) \qquad (2.10)$$

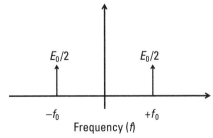

Figure 2.3 Spectrum of simple electric field.

$$e(t,R) = E_0 \cos(\omega_0 t - k_0 R) \qquad (2.11)$$

$$\omega_0 = 2\pi f_0 \qquad (2.12)$$

$$k_0 = \frac{2\pi}{\lambda_0} \qquad (2.13)$$

$$c = \frac{\omega_0}{k_0} = f_0 \cdot \lambda_0 \qquad (2.14)$$

The speed of light is c and the wavelength of the electric field is λ_0. The frequency represents the number of oscillations per unit of time (e.g., seconds) at a particular place. In the same way, the wavelength represents the number of oscillations per unit of distance (e.g., meters) at a particular time. At a distance R from the source of the radiation, the field is identical to the radiation at the origin except for a phase shift from the original field.

Radar frequencies are from several megahertz to tens of gigahertz. The speed of light is about 3×10^8 m/s. The radar wavelengths vary from hundreds of meters to millimeters. At the present time, typical ASM radar sensors are X-band, Ku-band, or millimeter wavelengths although other frequencies are used. Most ASM pulsed Doppler radar sensors use X-band, because of its ability to penetrate water vapor and the convenient size of the radar components. X-band wavelengths are of several centimeters. This is the typical frequency band used for many other popular applications such as weather radars. Thus, there are usually many conflicting systems generating radiations that tend to confuse the fleet defensive sensors. Table 2.1 shows the customary radar frequency designations.

The typical ASM radar contains a device that generates precise oscillating signals of various frequencies. This device serves as a reference clock for the system. This basic signal (f) is typically an oscillation at a low frequency (relative to RF). The engineer uses a variety of radar components to convert this oscillation to the oscillations of desired characteristics. For example, the basic signal is mixed with other signals such as a signal at a frequency designated the local oscillator frequency. This generates a signal consisting of sum and difference frequencies. This result is band-pass filtered to isolate one of the frequency components. Repeating this process in stages raises the signal frequency to the desired radar frequency (RF). Although these devices are nonlinear components, the result can be understood as a linear frequency shift [4, 5]. For example, assume that the original frequency is f. The signal

Table 2.1
Typical Radar Frequencies

Designation	Typical Frequency Range	ASM Sensors
HF	3 to 30 MHz	—
VHF	30 to 300 MHz	—
L	1 to 2 GHz	—
S	2 to 4 GHz	—
C	4 to 8 GHz	—
X	8 to GHz	Yes
Ku	12.5 to 18 GHz	Yes
K	18 to 27 GHz	—
Ka	27 to 40 GHz	—
Mm	40 to 300 GHz	Yes

frequency can be upshifted by mixing with another signal and filtering out the unwanted frequencies with the proper choice of components. This example is illustrated in Figure 2.4.

$$\cos(2\pi f_{LO} t) \cdot \cos(2\pi f t) \\ = \frac{1}{2} \cdot \left\{ \cos\left[2\pi(f_{LO} + f)t\right] + \cos\left[2\pi(f_{LO} - f)t\right] \right\} \quad (2.15)$$

The signal source and other components of the radar transmitter generate energy at the desired frequency and at other desired characteristics. The amplifier raises this signal to the desired amplitude. The signal is then coupled to an antenna and emitted from the seeker. The antenna enables the transition of the signal from the source electronic hardware to the atmosphere. (The antenna is considered in detail below.) In addition, the configuration of the antenna

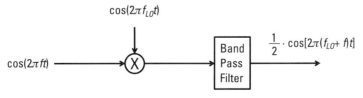

Figure 2.4 Radar mixer model.

determines the polarization of the radiation. Typical ASM peak amplitudes are of the order of hundreds of watts.

As described, this transmitter generates continuous wave (CW) energy. Suppose the radiation is interrupted by a switch such that a burst of energy is transmitted via the antenna for a time interval pulse width (PW). Figure 2.5 shows a simplistic representation of the radar transmitter system for CW radar and for pulsed radar.

The result of this switch is to constrain the transmit signal in time. The standard terminology is that a window function has truncated the CW signal in time. That is, the signal is multiplied by a (rectangular) window function $w(t)$ of PW.

$$w(t) = \frac{1}{\text{PW}} \qquad |t| \leq \frac{\text{PW}}{2} \qquad (2.16)$$

$$w(t) = 0 \qquad |t| > \frac{\text{PW}}{2} \qquad (2.17)$$

Using (2.5) for the inverse Fourier transform, the spectrum of this rectangular window function is the common sinc function. This function is qualitatively illustrated in Figure 2.6.

$$W(f) = \int e^{-2\pi i f t} w(t) \, dt \qquad (2.18)$$

$$W(f) = \frac{\sin(\pi f \text{PW})}{\pi f \text{PW}} \qquad (2.19)$$

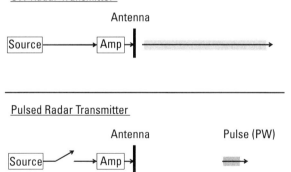

Figure 2.5 Simple radar transmitter systems.

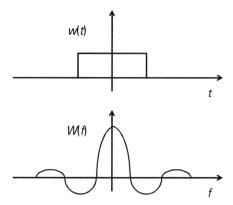

Figure 2.6 Spectrum of the rectangular window function.

As discussed, the spectral side lobes of this pulsed signal can be lowered by choosing a modified window function. This is typically known as multiplying the window function or the signal by a weighting function. Lowering the side lobes in this manner generally has the effect of broadening the spectral width of the main peak and lowering the level of the spectral peak.

The transmitted pulsed signal is now (still neglecting the polarization)

$$s(t) = e(t) \cdot w(t) = E_0 \cos(2\pi f_0 t) \cdot w(t) \quad (2.20)$$

The Fourier transform of the product of two functions is the convolution of the two transforms. Thus, the spectrum of the pulsed signal is

$$S(f) = \int e^{-2\pi i f t} e(t) \cdot w(t) \, dt = \int E(f - f') \cdot W(f') \, df' \quad (2.21)$$

The resulting spectrum is qualitatively illustrated in Figure 2.7.

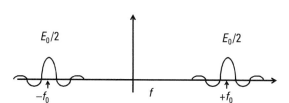

Figure 2.7 Spectrum of a transmitted electromagnetic pulse.

Again the function $S(f)$ represents the amplitude spectrum. A related function of interest is the power spectrum. The time correlation function is defined as

$$C'(\tau) = \int s(t+\tau) \cdot s^*(t)\, dt \quad (2.22)$$

From the above expressions and following [1, 2], one can relate the time correlation of the $s(t)$ function to the Fourier transform of the power spectrum $P(f)$ of $s(t)$ as follows:

$$C'(\tau) = \int (|S(f)|)^2 e^{2\pi i f \tau}\, df = \int P(f) e^{2\pi i f \tau}\, df \quad (2.23)$$

$$P(f) = (|S(f)|)^2 \quad (2.24)$$

The bandwidth (BW) of the power spectrum is the frequency width at the half-power points for a typically shaped power spectrum. The value of the correlation function at 0 time shift is the total power.

$$C'(0) = \int P(f)\, df = \int \left[|s(t)|\right]^2 dt \quad (2.25)$$

This correlation function is typically normalized to 1 at $\tau = 0$ as

$$C(\tau) = \frac{C'(\tau)}{C'(0)} \quad (2.26)$$

$$C(0) = 1 \quad (2.27)$$

From the defining expression (see [2.22]), it is seen that the value of the correlation function is larger when the shifted function is similar to the *unshifted* function. A function is said to be *decorrelated* if its normalized correlation function is smaller than a specified value after the decorrelation time τ_0. In this case, the shifted function is no longer similar to the unshifted function. Examining the correlation function, it is seen that the shorter the time for the time function to decorrelate, the wider the frequency BW of the corresponding power spectrum. For the case of a simple pulsed sinusoid (rectangular window)

$$\tau_0 = \frac{1}{BW} \quad (2.28)$$

Figure 2.8 illustrates the relationship between the amplitude and power spectra and the signal BW and the correlation time.

Consider a very simple example. Let

$$P(f) = P_0 \qquad |f| < \frac{BW}{2} \qquad (2.29)$$

$$P(f) = 0 \qquad |f| > \frac{BW}{2} \qquad (2.30)$$

Then

$$C(\tau) = \mathrm{sinc}(\pi BW \tau) \qquad (2.31)$$

For this example, the first zero of the correlation function occurs at

$$\tau = \frac{1}{BW} \qquad (2.32)$$

Figure 2.9 illustrates the correlation function for this simple example. The top figure is the correlation function and the power spectrum for a narrow bandwidth signal. The lower portion of the figure shows the power spectrum for a wider bandwidth signal. As illustrated by its correlation function and (2.31), this signal has a much shorter correlation time.

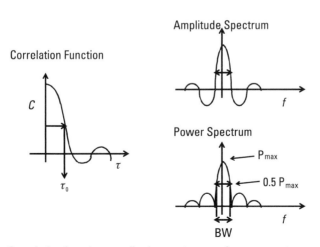

Figure 2.8 Correlation function, amplitude spectrum, and power spectrum.

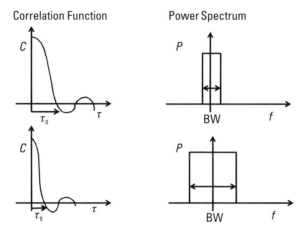

Figure 2.9 Correlation and simple spectrum.

Again consider the expression for the above correlation function in (2.23). Approximate the exponential function by the first several terms of its series expansion representation.

$$C'(\tau) = \int P(f) e^{2\pi i f \tau} df \approx \int P(f) \cdot \left(1 + 2\pi i f \tau - 2\pi^2 \tau^2 f^2\right) \cdot df \quad (2.33)$$

This expression relates the correlation function to the total power and the first and second moments of the power spectrum. The imaginary part is related to the mean frequency. The real part is related to the spectral BW.

$$C'(\tau) = C'(0) \cdot \left(1 - 2\pi^2 \tau^2 \overline{f^2} + i 2\pi \tau \overline{f}\right) \quad (2.34)$$

$$C(\tau) = \left(1 - 2\pi^2 \tau^2 \overline{f^2}\right) + i 2\pi \tau \overline{f} \quad (2.35)$$

For the spectrum described in (2.29) and (2.30), the approximation in agreement with (2.31) is

$$C(\tau) = \operatorname{sinc}(\pi \mathrm{BW} \tau) \approx 1 - \left(\frac{\pi^2}{3 \cdot 2}\right) \cdot (\mathrm{BW} \cdot \tau)^2 \quad (2.36)$$

For another example, consider the simple spectrum illustrated in Figure 2.10.

$$P(f) = P_0 \cdot \left(1 + \frac{f}{\mathrm{BW}}\right) \quad -\mathrm{BW} < f < 0 \quad (2.37)$$

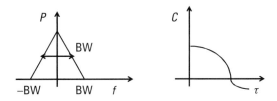

Figure 2.10 Simple spectrum.

$$P(f) = P_0 \cdot \left(1 - \frac{f}{BW}\right) \qquad 0 < f < BW \qquad (2.38)$$

$$P(f) = 0 \qquad |f| > BW \qquad (2.39)$$

The approximation for this case gives (for small τ)

$$C(\tau) \approx 1 - \left(\frac{\pi^2}{3}\right) \cdot (BW \cdot \tau)^2 \qquad (2.40)$$

A useful parameter for the EP analyses later in this work is the Lag-1 value of the normalized correlation function. This is an estimate of the normalized correlation function at one time sample (such as one coherent processing interval or one pulse repetition interval [PRI]). Define

$$T = \frac{1}{\text{Sample\_Rate}} \qquad (2.41)$$

$$\text{Lag} - 1 \text{ Value} = C(T) \qquad (2.42)$$

The following is a simple but useful implementation of an approximation for the Lag-1 value of the normalized correlation function. Define the following terms for discrete time samples (subscript n)

$$M0 = \frac{1}{N} \sum s_n \qquad (2.43)$$

$$S0 = \frac{1}{N} \sum s_n \cdot s_n \qquad (2.44)$$

$$S1 = \frac{1}{N}\sum s_n \cdot s_{n+1} \quad (2.45)$$

Estimates of each of these terms can be continually updated for each new sample (s_N) via simple low-pass filters with gain g, where the minus sign subscript indicates the previous time estimates

$$M0 = (1-g)s_N + gM0_- \quad (2.46)$$

$$S0 = (1-g)s_N \cdot s_N + gS0_- \quad (2.47)$$

$$S1 = (1-g)s_N \cdot s_{N+1} + gS1_- \quad (2.48)$$

Then, at any time, an estimate of the Lag-1 value is

$$C(T) = \frac{S1 - M0^2}{S0 - M0^2} \quad (2.49)$$

2.2 Dynamic Range and Gain Control

Repeating the above expression, the time dependence at the spatial origin of the electric field amplitude is (2.1) using f_c as the carrier frequency

$$e(t) = E_0 \cos(2\pi f_c t) \quad (2.50)$$

This can be written in complex form using Euler identities.

$$e(t) = \frac{E_0}{2}\left[e^{2\pi i f_c t} + e^{-2\pi i f_c t}\right] \quad (2.51)$$

The frequency representation (or spectrum) of a function of time is expressed via the Fourier transform and its inverse.

$$S(f) = \int e^{-2\pi i f t} s(t)\, dt \quad (2.52)$$

$$s(t) = \int e^{2\pi i f t} S(f)\, df \quad (2.53)$$

Using this representation for the simple electric field and using the impulse function, the spectrum of this signal is written as (2.54) and shown previously in Figure 2.3.

$$E(f) = \frac{E_0}{2}\delta(f - f_c) + \frac{E_0}{2}\delta(f + f_c) \qquad (2.54)$$

Assume that radar transmits the pulse of Figure 2.5 and (2.20). The transmitted pulsed signal is now

$$e_T(t) = e(t) \cdot w(t) = E_0 \cos(2\pi f_c t) \cdot w(t) \qquad (2.55)$$

The Fourier transform of the product of two functions is the convolution of the two transforms. Thus, the spectrum of the pulsed signal is

$$E_T(f) = \int e^{-2\pi i f t} e(t) \cdot w(t) \, dt = \int E(f - f') \cdot W(f') \, df' \qquad (2.56)$$

The resulting spectrum is illustrated in Figure 2.7. Consider now a slightly more general transmitted pulse, and for convenience neglect the rectangular window function. The more general transmitted pulse contains a simple modulated phase shift (ψ) added to the basic fixed frequency.

$$e(t) = E_0 \cos\{2\pi[f_c t + \psi(t)]\} \qquad (2.57)$$

Now assume that an echo of the transmitted pulse from an object at range R as shown in Figure 2.11 arrives back at the sensor antenna (at z = 0) after a time delay

$$\text{time delay} = \frac{2R}{c} \qquad (2.58)$$

$$e_1(t) = A_0 \cos\left\{2\pi[f_c t + \psi(t)] - 2\pi f_c \frac{2R}{c}\right\} \qquad (2.59)$$

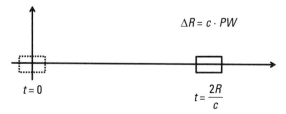

Figure 2.11 Reflected pulse.

$$e_1(t) = A_0 \cos\{2\pi[f_c t + \psi(t) - \varphi]\} \quad (2.60)$$

$$\varphi = \frac{2R}{\lambda_c} \quad (2.61)$$

The first thing that is normally done to the signal after passing from the antenna into the radar processing is the possibility of attenuating the signal to prevent saturation of any receiver components. This attenuation setting can be adaptively adjusted by an automatic gain control (AGC) algorithm based on prior information. The next action is again to mix the signal several times in stages to reduce the carrier frequency to a baseband frequency. As before, filters (band pass or low pass) are used to remove unwanted mixer components (see [2.15] and Figure 2.4). The received signal can be represented as

$$e_1(t) = A_0 \cos\{2\pi[(f_c - f_{LO})t + \psi(t) - \varphi]\} \quad (2.62)$$

$$f_0 = f_c - f_{LO} \quad (2.63)$$

Finally, it is assumed that this signal is corrupted by additive receiver white noise simply modeled as follows:

$$x(t) = A_0 \cos\{2\pi[f_0 t + \psi(t) - \varphi]\} + \sigma_0 n(t) \quad (2.64)$$

$$\langle n(t) \rangle = 0 \quad (2.65)$$

$$\langle n(t) n^*(t') \rangle = \delta(\tau - \tau') \quad (2.66)$$

Using the common identity for the cosine function equation (2.64), the signal portion can be written as

$$s(t) = A_0 \big(\cos(2\pi f_0 t)\cos\{2\pi[\psi - \varphi]\} - \sin(2\pi f_0 t)\sin\{2\pi[\psi - \varphi]\}\big) \quad (2.67)$$

This reflected signal can be analyzed and characterized via a quadrature detector as shown in Figure 2.12.

The received signal is multiplied by or mixed with the signals and then low-pass filtered (or averaged) as shown. The ability to properly resolve these signals in both amplitude and phase requires that the phase of the reference signal be relatable to the phase of the transmitted signals. This property is the

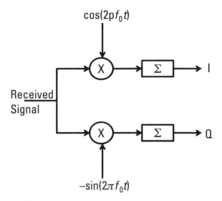

Figure 2.12 Quadrature detector.

coherent property of the radar. Being able to control the phase within a pulse is designated the intrapulse coherence. Being able to control the phase over multiple pulses is designated the interpulse coherence.

Using the low-pass filter over a sample time interval converts the RF pulse signal to a baseband signal that is detected as two output values. The first term on the right-hand side of the figure is said to be in phase with the original signal from the system master clock. Thus, the output from the top branch is called the *in-phase* component or *I*. The second term is *out of phase* with the original signal by $\pi/2$. The output from the lower branch is called the *quadrature* component or *Q*. Thus, the signal portion of the detector output sample is

$$I = A_0 \cos\left[2\pi(\psi - \varphi)\right] \qquad (2.68)$$

$$Q = A_0 \sin\left[2\pi(\psi - \varphi)\right] \qquad (2.69)$$

By Fourier analysis, these terms are independent. Since these two terms represent independent outputs, they can be combined as a single complex number representing this sample of the output signal plus noise at a particular sample time

$$\text{sample} = I + iQ = A_0 e^{2\pi i(\psi - \varphi)} + \sigma_0 n' \qquad (2.70)$$

Knowing the receiver characteristics, this complex number fully describes this time sample of the received signal plus receiver noise. Thus far the description has been overly simplified. The pieces described can be represented as follows.

As the signal enters the receiver, it may need to be attenuated to prevent it from saturating the later components. This may require implementation of an AGC circuit.

The receiver next consists of multiple stages of mixers and band-pass filters [3]. At each stage, the signal is mixed with a reference signal resulting in both sum and difference frequency components. The higher (sum) frequency components are then filtered (attenuated) by averaging or low pass filtering, allowing only the lower (difference) frequency component to pass through. A single representative stage in a simplified receiver is shown in Figure 2.13.

The effective bandwidth of the components controls the amount of receiver noise typically expressed as a reference thermal energy (kT_0) times the receiver BW and receiver noise figure. After the energy is passed through the final receiver components, the resultant is integrated and both the I and Q [4, 5] components are evaluated as digital numbers via ADCs. This noise level may be further adjusted via another attenuator just before splitting the signal to control the noise level out of the quadrature detector. The purpose of this attenuator is to set the receiver noise level to a desired number for processing after the ADC. For example, the data may be adjusted such that the receiver noise level is impacting the lowest one or two bits in the ADC. As mentioned above, an initial attenuator may be used to keep the I and Q components from saturating the various analog receiver components and especially the ADCs. Figure 2.14 represents the analog receiver levels on the left-hand side and the resultant corresponding digital levels on the right-hand side.

As shown above, the ADC generates digital numbers that represent the sample (I and Q) of received energy plus receiver noise at a particular time

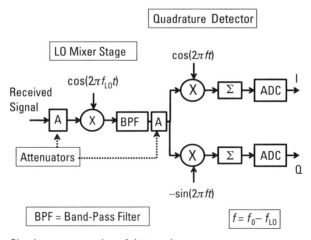

Figure 2.13 Simple representation of the receiver.

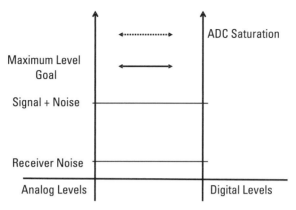

Figure 2.14 Analog-to-digital conversion.

sample. Assuming that the energy is from an echo of the transmitted pulse plus jamming, this sample represents information about a possible reflecting object at a particular range cell. To maintain fidelity of the information requires the information to be within the dynamic range of the ADC. Assuming the ADC contains b bits, the dynamic range in decibel is approximately

$$\text{DR} = 10 \cdot \log\left(\frac{P_{max}}{P_{min}}\right) \approx 6b \text{ dB} \tag{2.71}$$

As noted above, an AGC algorithm may be used to lessen the possibility of data in saturation. The AGC may not be required for a well-designed receiver, since the ADC of modern ASM LPI radar typically has 12 or more bits resulting in a dynamic range of better than 72 dB. Some room must be allowed to keep the signals from clipping. There are many excellent references [3–5] on radar receivers. The main interest of this work is in the digital EP techniques in the digital processors that follow the radar receivers.

First, the samples of interest must be detected. In classic radar, the echo is a blip on an A scope. In modern ASM radar, the detection of samples of interest is the result of matched filter processing. Consider a simple signal plus noise

$$x(t) = a \cdot s(t) + \sigma \cdot n(t) \tag{2.72}$$

The output of the matched filter (h) at time τ is defined as

$$\chi(\tau) = \int x(t) \cdot h(\tau - t)\, dt \tag{2.73}$$

This output has a signal part and a noise part from (2.72). The noise part is a statistical variable with mean zero and variance (assuming an arbitrary scaling factor for the filter)

$$\langle \chi(\tau)\chi^*(\tau') \rangle = \sigma^2 \int h(\tau - t) h^*(\tau' - t) \, dt \qquad (2.74)$$

$$\langle \chi(\tau)\chi^*(\tau') \rangle = \sigma^2 \int e^{2\pi i f(\tau-\tau')} \left[|H(f)|\right]^2 df \qquad (2.75)$$

$$\langle \chi(\tau)\chi^*(\tau') \rangle = \sigma^2 C_h(\tau - \tau') \qquad (2.76)$$

Consider the signal part at a particular time corresponding to the arrival of the echo at t_0. The signal part is

$$\chi(t_0) = a \int e^{2\pi i f t_0} S(f) H(f) \, df \qquad (2.77)$$

The detection theory result is that the optimal detection occurs at the highest signal-to-noise ratio (SNR). Thus, we want to maximize the magnitude squared of the signal part of the matched filter output, that is, (2.77). The Schwarz inequality is

$$\left[\left|\int A(f) B(f) \, df\right|\right]^2 \leq \int \left[|A(f)|\right]^2 df \int \left[|B(f)|\right]^2 df \qquad (2.78)$$

The two sides of this expression are equal if and only if

$$B(f) = A^*(f) \qquad (2.79)$$

Thus, the magnitude squared of the signal portion of the output of the matched filter is maximum at t_0 when

$$H(f) = e^{-2\pi i f t_0} S^*(f) \qquad (2.80)$$

$$h(t) = s^*(t_0 - t) \qquad (2.81)$$

Repeating (2.73) with the substitution of (2.81)

$$\chi(\tau) = \int x(t) \cdot s^*\left[t - (\tau - t_0)\right] dt \qquad (2.82)$$

$$\chi(\tau) = a \cdot C_s(\tau - t_0) + \chi_n(\tau) \quad (2.83)$$

It is noted from expressions (2.76) and (2.83) that the SNR of the matched filter output is

$$\mathrm{SNR} = \frac{a^2}{\sigma^2} \cdot C_s(0) = \frac{a^2}{\sigma^2} \quad (2.84)$$

Thus, the SNR or the detection capability depends only on the total power in the echo. It does not depend on the details of the phase modulation within the pulse. This is an important result for LPI radar.

The interpretation of (2.82) and (2.83) is that the signal portion of the matched filter output is the signal level times the normalized autocorrelation function of the transmitted pulse and peaks at the echo delay time. The output of the matched filter has additive noise with statistics described above, that is, mean zero and variance given by (2.76). This time delay is a measure of the range to the object by (2.58). From the correlation function discussion above, the resolution capability of the range filter is governed by the BW of the filter or correspondingly by the bandwidth of the transmitted pulse.

$$\delta R = \frac{c}{2 \cdot \mathrm{BW}} \quad (2.85)$$

Figure 2.15 is a representation of the transmit pulse and the echo for a simple rectangular pulse with no phase modulation. The lower portion of the figure represents the output of the matched filter at high SNR for this simple pulse.

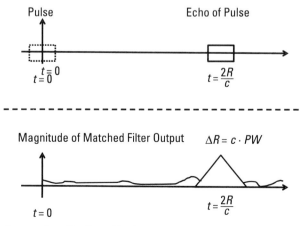

Figure 2.15 Range of a reflecting object.

It is not practical to calculate integrals over infinite time or to implement algorithms for continuous signals. In analogy with (2.52) and (2.53), the discrete Fourier transform (DFT) is defined. For the receiver example above (Figure 2.13), an ADC samples the time series of (2.50) at a sample rate (SR), taking N samples over a time interval T.

$$\text{SR} = \delta t^{-1} \tag{2.86}$$

$$N = \text{SR} \cdot T = \frac{T}{\delta t} \tag{2.87}$$

The digitized expression for the time variation of the field is N values

$$e(n) = E_0 \cos(2\pi f_0 \delta t n) = E_0 \cos\left(2\pi f_0 T \frac{n}{N}\right) \tag{2.88}$$

Suppose there is a signal that is identical to $e(t)$ but at a frequency SR greater than f_0

$$f = f_0 + \text{SR} \tag{2.89}$$

$$\tilde{e}(n) = E_0 \cos\left(2\pi f T \frac{n}{N}\right) = E_0 \cos\left[2\pi (f_0 + \text{SR})\frac{nT}{N}\right] \tag{2.90}$$

With some simple identities and (2.53)

$$\tilde{e}(n) = E_0 \left[\cos\left(2\pi f_0 T \frac{n}{N}\right)\cos\left(2\pi \text{SR} \frac{nT}{N}\right) \right. \\ \left. - \sin\left(2\pi f_0 T \frac{n}{N}\right)\sin\left(2\pi \text{SR} \frac{nT}{N}\right)\right] \tag{2.91}$$

$$\tilde{e}(n) = e(n) \tag{2.92}$$

The interpretation of this simple result is that the expression for the spectrum of the sampled time series repeats every SR. Thus, it is sufficient to calculate the spectral values in a frequency band SR wide such as [0, SR] or [−SR/2, SR/2].

The expression for the DFT of the N time series samples is [4]

$$S(k) = \sum s(n) e^{\frac{-2\pi i k n}{N}} \tag{2.93}$$

$$n \in [0, N-1] \qquad (2.94)$$

The inverse DFT is

$$s(n) = \frac{1}{N} \sum S(k) e^{\frac{2\pi i k n}{N}} \qquad (2.95)$$

$$k \in [0, N-1] \qquad (2.96)$$

This can be verified by substituting the expression (2.93) into (2.95) and using the result

$$\frac{1}{N} \sum e^{\frac{2\pi i k(n-n')}{N}} = \frac{1}{N} e^{\frac{\pi i (n-n')(N-1)}{N}} \cdot \frac{\sin(\pi[n-n'])}{\sin\left(\frac{\pi[n-n']}{N}\right)} \begin{matrix} = 1 \text{ if } n = n' \\ \\ = 0 \text{ if } n \neq n' \end{matrix} \qquad (2.97)$$

It is noted that the DFT is defined at the N frequencies

$$f = k\frac{\text{SR}}{N} = k\frac{1}{T} \qquad (2.98)$$

The N frequencies are spaced at $1/T$. Again, the N frequencies span a band SR wide (see [2.86]). The time series is both digitized and effectively multiplied by a window function of width T. Thus, the underlying spectrum is the convolution of the spectrum of the underlying time function and the window function. The filter response of the DFT is the set of N sinc functions of bandwidth $1/T$. Each of the sampled filters is centered at the zeroes of the other filters. Consider the DFT of a pure complex sinusoid function

$$s(n) = e^{\frac{2\pi i f T n}{N}} \qquad (2.99)$$

Then, the DFT in filter k is

$$S(k) = e^{\frac{\pi i (fT-k)(N-1)}{N}} \cdot \frac{\sin[\pi(fT-k)]}{\sin\left[\frac{\pi(fT-k)}{N}\right]} \qquad (2.100)$$

For filter values k close to fT

$$S(k) \approx e^{\frac{\pi i (fT-k)(N-1)}{N}} \cdot N \cdot \text{sinc}[\pi(fT-k)] \quad (2.101)$$

The interpretation of this result is shown in Figure 2.16 for a time series with frequency centered on one of the DFT filters.

$$fT = k_0 \quad (2.102)$$

The upper portion of the figure shows the location of the sinusoid frequency and several of the nearby filters for k values close to fT. Since the frequency is at the same location as one of the filters, the response is a maximum in that filter and zero in all the other filters.

In Figure 2.17, the frequency is not centered in a filter. In this case, there is some response in each of the filters because of the DFT filter side lobes.

Also, the highest filter response is somewhat reduced from the actual value of the time series amplitude because of this frequency location mismatch. The location of the underlying frequency and the signal amplitude are typically estimated by interpolation. Today, there are inexpensive DSP processors containing fast DFT algorithms. The engineer must send the array of data to the proper memory location and specify the number of data samples in the time series array.

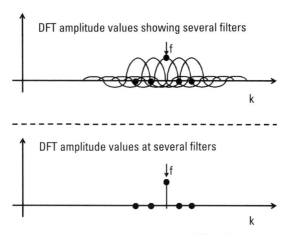

Figure 2.16 DFT of a complex sinusoid centered at a DFT value.

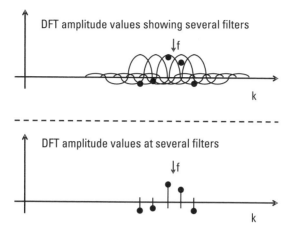

Figure 2.17 DFT of a complex sinusoid not centered at a DFT value.

This reduction in peak value is commonly designated the DFT filter straddle loss. Another common way to reduce the possible straddle loss is to *zero fill* the time series array. For example, if the N point time series is doubled in length by adding an array of N zeroes, the DFT has $2N$ filter values. The filter response is exactly the same as above, since the integration time is still T. The effect of the zero fill is to position N additional filters between each of the previous filters. Figure 2.18 illustrates the result.

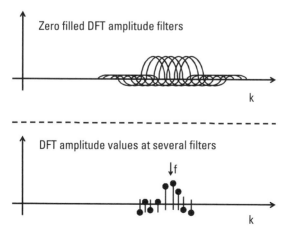

Figure 2.18 Effect of zero fill on the DFT.

2.3 Coherent Gain and Noncoherent Gain

The engineer must understand the basic concept of digital processing gain. Consider a set of N identical complex samples

$$S = \{s_n | s_n = A_0 e^{2\pi i \varphi_s}\} \qquad (2.103)$$

If these N identical samples are combined (added) the resulting variable is a constant complex number. The value and power (magnitude of the vector squared) are

$$\text{sum}\_s_T = \sum s_n = N A_0 e^{2\pi i \varphi_s} \qquad (2.104)$$

$$\left(|\text{sum}\_s_T|\right)^2 = N^2 \cdot A_0^2 \qquad (2.105)$$

These are the characteristics of coherent gain. The magnitude of the sum increases as N, and the magnitude squared increases as N^2.

In this work, a simple model of noise samples shall be used to demonstrate important features of the ASM sensor. Assume noise-like samples with fixed amplitude but with random phase in sample index n

$$\text{NSE} = \{\text{nse}_n | \text{nse}_n = \sigma e^{2\pi i \varphi_n}\} \qquad (2.106)$$

The sum of the N terms is again a random variable. Examining the expectation value and the variance (power) gives

$$\langle \text{sum}\_n_T \rangle = \left\langle \sum \text{nse}_n \right\rangle = 0 \qquad (2.107)$$

$$\left\langle \left(|\text{sum}\_n_T|\right)^2 \right\rangle = \left\langle \sum \text{nse}_n \cdot \sum \text{nse}_n^* \right\rangle = N \cdot \sigma^2 \qquad (2.108)$$

These are the characteristics of noncoherent gain. The expectation value of the sum is zero. The expected noise power (variance) increases by N. These features are illustrated in Figure 2.19.

In the same manner, the ratio of the coherent signal total power to the noise power or SNR increases as the number of terms in the combination (N)

$$\text{SNR} = \frac{N^2}{N} \cdot \frac{A_0^2}{\sigma^2} = N \cdot \text{SNR}_0 \qquad (2.109)$$

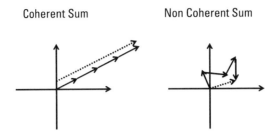

Figure 2.19 Concepts of coherent gain and noncoherent gain.

Table 2.2 summarizes these important properties of coherent integration gain and noncoherent integration gain.

Now, the powerful concept of radar range compression can be understood as the coherent combination of multiple range (or time) samples. This is a filtering option resulting in a much improved range estimate with consequent range filter side lobes. As noted above, the echo of a transmitted pulse from an object at range R arrives at the sensor after a time delay

$$\text{time delay} = \frac{2R}{c} \qquad (2.110)$$

The capability to resolve the estimate of this range is given by the well-known formula related to the bandwidth of the pulsed signal given above and is repeated here

$$\delta R = \frac{c}{2 \cdot \text{BW}} \qquad (2.111)$$

This expression reveals a significant advantage of the coherent radar or the LPI radar. For a simple pulse, BW is the inverse of PW (BW = 1/PW). Thus, to achieve the finest range resolution requires the narrowest pulse possible. However, the ability to detect a target depends on the amount of reflected energy received. This requires transmission of more energy in the pulse. To

Table 2.2
Gain Properties Summary

	Coherent	Noncoherent
Expected Value of Sum	N	0
Expected Value of Sum Squared	N^2	N

transmit more energy in the pulse requires using a high peak power in the narrow pulse, which is somewhat easier for the target ship to detect. For a simple pulse, the two contradictory requirements require using a very high transmitter energy that is more easily detectable or using a wider (in time) pulse that corresponds to losing range resolution.

Suppose the transmitted pulse is a contiguous sequence of J very narrow (in time) subpulses of pulse width PW_J at low peak power. Each subpulse has excellent range resolution, but very low energy and less detection capability. Note that each subpulse is delayed in time the same amount relative to its corresponding transmitted subpulse. If the initial phase of each transmitted pulse is coded, then each subpulse echo contains this coded phase. Assume that digital samples are collected as range samples corresponding to the subpulse range resolution. For each subpulse (designating the subpulse as number j), the J samples via the quadrature detector are the J contiguous time samples, each at the same range relative to the transmit time of each subpulse

$$I_j + iQ_j = e^{2\pi i \varphi_j} \cdot A_0 e^{2\pi i \varphi_s} \tag{2.112}$$

The range matched filter described above can be implemented since the phase code is known by the radar processor. The signal matched filter is the phase code of the transmit pulse. The matched filter contains J elements

$$h(-t) = s^*(t) = \left\{ e^{-2\pi i \varphi_j} \mid j = 0, J-1 \right\} \tag{2.113}$$

The process of matched filtering is to multiply each subpulse sample by this known filter and then sum (coherently) over the subpulse number j. The coherent sum of range samples is

$$\text{Range Sample} = \sum e^{-2\pi i \varphi_j} \cdot \left(I_j + iQ_j \right) = J \cdot A_0 e^{2\pi i \varphi_s} \tag{2.114}$$

By transmitting a wide pulse that is thus modulated (or coded), one can retain the range resolution of the very narrow pulse while increasing the total energy in the compressed range sample by coherent integration. This is the LPI concept of pulse compression [5–7].

$$PW = J \cdot PW_J \tag{2.115}$$

$$\delta R = c \cdot \frac{PW_J}{2} \tag{2.116}$$

$$\text{Energy} = J \cdot A_0 \quad (2.117)$$

$$\text{Power} = J^2 \cdot A_0^2 \quad (2.118)$$

This makes possible high SNR matched filter output with fine range resolution via a low peak energy but wide (in time) pulse. This is a powerful EP technique. Practical examples of this range compression technique are described in detail below. The top portion of Figure 2.20 shows the simple received echo pulse. The lower portion illustrates the matched filter output principal showing the peak value and several arbitrary range filter side lobes. Also, effectively the output of the pulse compression filter has converted the time samples to fully integrated range samples.

These pulse compression techniques lead to all of the standard filter response issues, including range side lobes and straddle loss. These range side lobes can be diminished using selected filter weighting functions. As noted above, the technique of pulse compression is said to exploit intrapulse coherency, since the coherent properties are within a single pulse.

As another simple example, assume the transmit pulse contains a five-chip Barker code. Barker codes are special codes of various lengths selected because the side lobe values are all either zero of one chip in magnitude. These codes are available and documented in [7]. The five subpulses are multiplied by the sequence

$$+1 \quad +1 \quad +1 \quad -1 \quad +1 \quad (2.119)$$

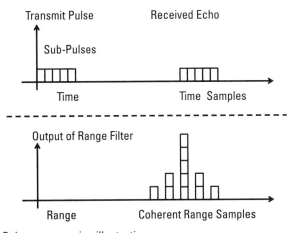

Figure 2.20 Pulse compression illustration.

This code can be represented as above

$$h(-t) = s^*(t) = \left\{ e^{-2\pi i \varphi_j} \middle| e^{-2\pi i 0}, e^{-2\pi i 0}, e^{-2\pi i 0}, e^{-2\pi i \frac{1}{2}}, e^{-2\pi i 0} \right\} \quad (2.120)$$

Figure 2.21 illustrates the reference clock signal and the corresponding transmitted pulse with the 5-chip Barker code imposed on it. Figure 2.22 illustrates the received echo time samples on the lower line as a representation of the output of the range matched filter.

The phase could be modulated with any of many known codes, including random phase shifts. The engineer can choose the code that meets any of several design criteria [7].

Another common pulse compression code is frequency modulation. Again the frequency could be varied randomly, but a common code is linear frequency modulation (LFM). Rather than making several discrete phase changes within the pulse, the phase is continuously modulated with time. From (2.57), consider the following transmit pulse:

$$e(t) = E_0 \cos\left\{2\pi\left[f_c t + \psi(t)\right]\right\} \cdot w(t) \quad (2.121)$$

For an LFM pulse

$$\psi(t) = \frac{mt^2}{2} \quad (2.122)$$

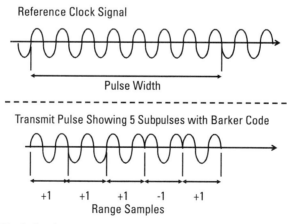

Figure 2.21 Clock signal representation and Barker coded transmit pulse.

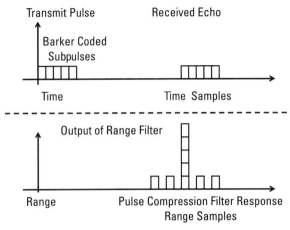

Figure 2.22 Five-chip Barker code pulse compression example.

Using the relationship that the instantaneous frequency is the time derivative of the phase, it is seen that the frequency varies linearly with time.

$$f(t) = f_c + mt \qquad \frac{-\text{PW}}{2} \leq t \leq \frac{\text{PW}}{2} \qquad (2.123)$$

$$f\left(\frac{-\text{PW}}{2}\right) = f_c - \frac{m \cdot \text{PW}}{2} \qquad (2.124)$$

$$f\left(\frac{\text{PW}}{2}\right) = f_c + \frac{m \cdot \text{PW}}{2} \qquad (2.125)$$

$$\text{BW} = m \cdot \text{PW} \qquad (2.126)$$

Typical ASM PWs are tens of microseconds. Typical bandwidths are tens of megahertz corresponding to range resolutions of 10 to 40m.

Consider a single LPI pulse. After the transmit pulse, a time passes corresponding to a particular range by the time required for the echo to return. Starting at this time, a swath of samples is processed via the radar receiver and sampled via the ADCs. This swath of samples covers from a minimum range to a maximum range of interest. These time samples are considered range samples by counting time from the transmit pulse time. Assuming possible range compression processing, the result is an array of range samples with range resolution given above. The same can be done for a second transmit pulse and so forth. Over a set of pulses, a two-dimensional array of samples can be collected. This two-dimensional array consists of range samples on one axis and corresponding pulse number or transmit pulse time on the other axis. The data array is illustrated in Figure 2.23.

2.4 Antenna

As mentioned above, another important component of the ASM seeker sensor is the antenna. To transition the RF energy pulse from the transmitter to the environment and to transition the echo RF energy from the environment back into the receiver requires an antenna. ASM sensors have many varieties of antennas such as a simple parabolic reflector, a twist Cassegrain reflector, a flat-plate array of dipole elements, or a variety of compound antennas. These antennas have a variety of polarizations, but the most common ASM antenna polarization is linear. There are many works dealing with the special topic of radar antennas.

The underlying purpose of the antenna is to achieve more gain in a particular direction by forming an antenna beam pattern. If the radiation is transmitted uniformly in all directions, the power is reduced with range by the area of the sphere.

$$P(R) = \frac{P_0}{4\pi R^2} \quad (2.127)$$

If the reflection from an object returns in the same manner, the power is reduced by the area again. The variable σ represents the amount of power returned from the reflector in the direction of the radar.

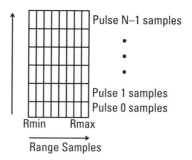

Figure 2.23 Range versus transmit pulse time array of samples.

$$P_{echo} = \frac{P_0 \sigma}{(4\pi)^2 R^4} \qquad (2.128)$$

Using an antenna, the transmit beam is focused and increased in a particular direction by the antenna gain function. In addition, the radiation received intercepts an effective area that is related to the antenna gain by the formula

$$G = \frac{4\pi}{\lambda^2} A_e \qquad (2.129)$$

Thus the formulas for the transmit power at range R and the subsequent reflected power are

$$P(R) = \frac{P_0 G}{4\pi R^2} \qquad (2.130)$$

$$P_{echo} = \frac{P_0 G^2 \lambda^2 \sigma}{(4\pi)^3 R^4} \qquad (2.131)$$

The ASM antenna is typically steered relative to the ASM body to achieve increased gain in a particular direction. In the future the antenna may be electronically steered, but at the present time most ASM antennas are mechanically steered.

There are several reasons for antenna steering. In the initial acquisition mode or a subsequent reacquisition mode after flying below the radar horizon for a period of time, the antenna is typically swept over an extant of angles in front of the ASM to search for targets. It was already mentioned that detection performance improves with increased SNR. Once a target is chosen the antenna is typically pointed in the direction of the target of interest to better estimate guidance corrections. The accuracy of a measurement typically improves with increased SNR. Steering may be especially required if the ASM is conducting deliberate maneuvers for mitigating the effectiveness of kinetic (anti-ASM) weapons. Finally, when tracking a target, some ASM seekers will deliberately jog the antenna slightly at times, to mitigate possible EA techniques. This is termed track while scan.

In this work, it is assumed that the model antenna is a flat plate consisting of a two-dimensional array of dipole subantennas. This array pattern is generally complicated and results from antenna design and modeling. To gain an understanding of the general antenna features for the purpose of this work, assume the array is a one-dimensional array of equally spaced dipole elements. Consider a radar echo from an object at a range much greater than the dimensions of the antenna. Each of the dipoles is a simple antenna element that can be coherently combined for greater gain. Due to its position in the array the return at each element is shifted in phase by an amount that varies with the angle relative to bore sight based on its relative location in the array. This leads to exactly the same expressions as above for DFT of the rectangular time window function in which the time sample parameter is now replaced by the element spatial location and the PW is replaced by the antenna size. For reasons that will become clear, assume that the upper half of the array and the lower half of the array are combined separately as indicated in Figure 2.24.

For an echo at long range R and at offset angle ψ, the path to each element is greater than the path to the next element by

$$\delta R = \delta d \cdot \sin \psi \qquad (2.132)$$

Letting φ correspond to the offset phase because of the reference range, the echo at an element n of the upper subantenna in the figure is represented as

$$\text{Upper\_Echo}_n = A \cdot \cos\left(2\pi ft - 2\pi\varphi + 2\pi n \frac{\delta R}{\lambda}\right) \qquad (2.133)$$

The echo at an element of the corresponding lower subantenna is

$$\text{Lower\_Echo}_n = A \cdot \cos\left(2\pi ft - 2\pi\varphi - 2\pi n \frac{\delta R}{\lambda}\right) \qquad (2.134)$$

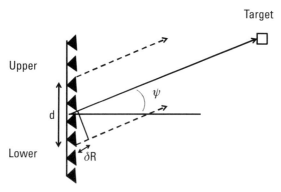

Figure 2.24 Antenna elements.

After these signals are processed in the receivers as above, they are

$$\text{Upper\_Echo}_n = A \cdot e^{(2\pi i f t - 2\pi i \varphi)} e^{+2\pi i n \frac{\delta R}{\lambda}} \quad (2.135)$$

$$\text{Lower\_Echo}_n = A \cdot e^{(2\pi i f t - 2\pi i \varphi)} e^{-2\pi i n \frac{\delta R}{\lambda}} \quad (2.136)$$

The return from each of these elements travels through the seeker receiver subsystem and is processed in the same manner. Figure 2.24 shows the spacing between the upper and lower antenna as d. With the assumptions given, d is the dimension of the two subantennas and

$$d = N \cdot \delta d \quad (2.137)$$

The sums of these signals over all antenna elements are

$$\text{Upper\_Echo} = A \cdot e^{(2\pi i f t - 2\pi i \varphi)} \cdot e^{+2\pi i \frac{d \sin \psi}{\lambda}} \cdot N \operatorname{sinc}\left(\frac{\pi d \sin \psi}{\lambda}\right) \quad (2.138)$$

$$\text{Lower\_Echo} = A \cdot e^{(2\pi i f t - 2\pi i \varphi)} \cdot e^{-2\pi i \frac{d \sin \psi}{\lambda}} \cdot N \operatorname{sinc}\left(\frac{\pi d \sin \psi}{\lambda}\right) \quad (2.139)$$

Typically, the antenna subelements are summed in the antenna at RF, and then the RF energy is combined in the antenna as a Sum and a Delta antenna. In this case, the total outputs of the receivers are

$$\text{Sum\_Echo} = 2 \cdot A \cdot e^{(2\pi i f t - 2\pi i \varphi)} \\ \cdot \cos\left(2\pi \frac{d \sin \psi}{\lambda}\right) \cdot N \operatorname{sinc}\left(\frac{\pi d \sin \psi}{\lambda}\right) \quad (2.140)$$

$$\text{Delta\_Echo} = 2i \cdot A \cdot e^{(2\pi i f t - 2\pi i \varphi)} \\ \cdot \sin\left(2\pi \frac{d \sin \psi}{\lambda}\right) \cdot N \operatorname{sinc}\left(\frac{\pi d \sin \psi}{\lambda}\right) \quad (2.141)$$

This sinc function beam pattern in angle has significant side lobes in angle. As mentioned above in the discussion of the range filter, a window weighting function can be applied to the antenna elements via slight spatial adjustments and/or amplitude gains to modify the antenna patterns. Antenna side lobes can be reduced at the expense of a slight loss of gain and a slight broadening of the beam patterns.

In addition, consider the echo received via the Delta antenna normalized by the echo received via the Sum antenna. Assuming small angle approximations

$$\frac{\text{Delta\_Echo}}{\text{Sum\_Echo}} = i \cdot \tan\left(2\pi \frac{d \sin \psi}{\lambda}\right) \approx i \cdot 2\pi \frac{d}{\lambda} \cdot \psi \qquad (2.142)$$

The imaginary part of the signal portion of this ratio is proportional to the angle off bore sight of the target. This is the basic concept of the monopulse angle measurement discussed earlier.

Assume in general that the antenna consists of four identical subantennas comprising the four spatial quadrants. These four subantennas are typically combined at RF to form a sum antenna, a left antenna, a right antenna, an upper antenna, and a lower antenna. These last four subantennas are then combined to form the elevation difference and the azimuth difference antenna. Thus far the discussion has been about the received antenna patterns. With isolation between the receivers and the transmitter the same antenna is generally used to transmit the radar pulses via the Sum antenna combination for maximum gain in the antenna pointing direction. Everything said about the Sum receive antenna patterns applies to the transmit antenna patterns by the principle of reciprocity.

In the case of a parabolic reflector antenna with horn feed, the antenna polarization is defined to be the same as the polarization of the feed. This is the polarization along bore sight of the antenna. However, this dominant polarization of the feed is complicated by a complementary beam component of orthogonal polarization generated by the shape of the parabolic reflector and the consequence of Maxwell's equations. Figure 2.25 illustrates the main polarization transmit pattern of a typical parabolic antenna. Figure 2.26 illustrates the full polarization component patterns generated for a simple parabolic antenna. At the peak, the orthogonal patterns (called Condon Lobes) are generally tens of decibels weaker than the peak of the main polarization pattern. In the ideal, the cross-polarization patterns are null along the cardinal (symmetry) axes.

It is important to note that the ASM antenna must view through an aerodynamically shaped dielectric radome as indicated in Figure 2.27. As stated in several works, the aerodynamic radome generally induces beams at polarizations and characteristics similar to that of the standard parabolic antenna even for the flat plate antenna array of dipoles [6–9]. In Figures 2.25 and 2.26, the Σ channel beam pattern (both transmit and receive) is shown

Pulsed Doppler Radar Basics 67

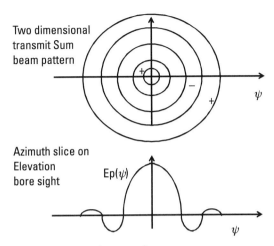

Figure 2.25 Parabolic antenna main transmit pattern.

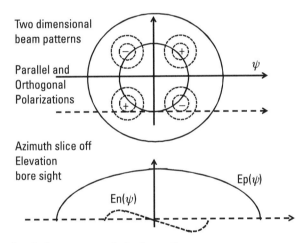

Figure 2.26 Parabolic antenna polarization patterns.

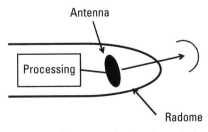

Figure 2.27 Mechanically steered antenna viewing through the radome.

as seen through the radome. The main beam is shown in both the nominal or parallel polarization and the four Condon-like lobes of the beam in the orthogonal polarization. Again, this is a consequence of the geometry, the radome, and Maxwell's equations.

In most of this work, only the two azimuth antennas are considered for simplicity. It is assumed that the transmitter radiates through the Σ antenna. The RF echoes are combined in the left and right antennas. The left and right antenna outputs are combined via RF hybrids as a Σ signal and a Δ signal. The two independent RF echo signals are then processed via identical and separate radar receivers. This generates two separate and statistically independent arrays of digital data as described above.

Note again that while antenna hardware has greatly improved over the years, it is very difficult to create and maintain identical RF receivers. For this reason, it is typical that a variety of internal calibration schemes are operated in the seconds prior to launch or periodically during the flight to measure various receiver unbalances. The results of these calibration algorithms are applied as the appropriate digital corrections during signal processing.

2.5 Doppler Effect

As shown in Figure 2.23, the ASM pulsed Doppler radar sensor transmits a sequence of pulses with PRI of T. After each transmitted pulse (p), a selected interval of digital time samples or range samples is processed in the receivers and stored. The time between the pulses is much greater than the PW and greater than the length (in time) of the range swath. These range samples for a particular pulse may have been combined via range compression processing. These samples contain information about any targets in this range swath.

The next task is to examine the time history of the samples in a particular range cell over the coherent processing interval (CPI) of P pulses (CPI = PT) [3, 4]. While the figure shows the array as a two-dimensional array, its memory structure is not useful. As the data is stored, it is usually manipulated by a device called a corner turn memory. Whereas the original data was collected with each range sample for a particular pulse being contiguous, it is stored in a manner in which the time samples at a particular range cell are contiguous as illustrated in Figure 2.28.

Consider a single range cell. The radar ASM platform is moving at high speed (v = about Mach 1 or faster). The target is moving at a much slower speed (v_T = 30 Kts or less). As a result, the range between the ASM and the target

Pulsed Doppler Radar Basics 69

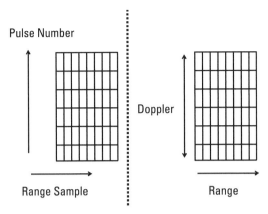

Figure 2.28 Data array example for single receiver channel.

is continually changing. Figure 2.29 (not drawn to scale) presents a snapshot of the relevant geometrical parameters.

The expressions above for the Σ and Δ channels are repeated

$$\text{Sum\_Echo} = 2 \cdot A \cdot e^{\left(2\pi i f_T pT - 2\pi i \frac{R(p)}{\lambda}\right)}$$
$$\cdot \cos\left(2\pi \frac{d \sin\psi}{\lambda}\right) \cdot N \operatorname{sinc}\left(\frac{\pi d \sin\psi}{\lambda}\right) \quad (2.143)$$

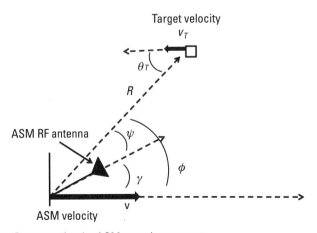

Figure 2.29 Geometry for the ASM pursuing a target.

$$\text{Delta\_Echo} = 2i \cdot A \cdot e^{\left(2\pi i f_T pT - 2\pi i \frac{R(p)}{\lambda}\right)}$$
$$\cdot \sin\left(2\pi \frac{d\sin\psi}{\lambda}\right) \cdot N\operatorname{sinc}\left(\frac{\pi d \sin\psi}{\lambda}\right) \quad (2.144)$$

$$R_m(p) \approx R_m - \left(v_T \cos\theta_T\right)pT - (v\cos\varphi)pT \quad (2.145)$$

Any target in an observed range cell is subject to a small change of range from pulse to pulse. Since the PRI of T is of the order of milliseconds, this change of range is typically much smaller than a range cell which is of the order of 10 to 30m [7]. So the target stays within the range cell during the CPI. Equation (2.145) is the first-order (linear in time) approximation to the range.

As indicated by this substitution into (2.143) and (2.144), the changing range with time corresponds to a linearly changing phase with time. This corresponds to a shift of the echo frequency during the observation interval. This frequency shift is the Doppler shift approximation for moving objects. This frequency for a target in the range cell can be measured via a conventional DFT or may include a window function to reduce the standard DFT filter side lobes. The Doppler processing converts the pulse time versus range sample array to a Doppler measurement versus range array.

$$\text{Output}\left(f_K, R_m\right) = \sum e^{-2\pi i f_K p} \cdot \text{sample}\left(p, R_m\right) \cdot \text{window}(p) \quad (2.146)$$

The Doppler processing is referred to as interpulse coherency because it exploits the coherency of the radar from pulse to pulse within the CPI. The radar reference clock must maintain good time stability over time periods greater than the CPI. This same processing is done for each of the two receivers in the radar corresponding to the antenna Σ output and the antenna Δ output.

A rigorous discussion of RF Doppler shift is available in many works [5]. The ASM processing may be implemented in a number of ways. It is generally beneficial in the radar implementation to set the frequency of the transmit pulses such that the frequency of the received echoes from a target with no radial motion relative to the sea surface, that is, a naval ship viewed broadside, when at the center of the antenna has a desired frequency in the receivers. If the desired received frequency is f_0, set the transmit frequency to be f_T

$$f_T = f_0 + \delta f \quad (2.147)$$

Using this transmitter pulse frequency (f_T), the frequency in the receivers of the echo from the target shown in Figure 2.29 is by (2.143) to (2.145)

Pulsed Doppler Radar Basics

$$f_R = f_T + f_T \frac{v}{c}\cos\varphi + f_T \frac{v_T}{c}\cos\theta_T \qquad (2.148)$$

During receiver processing the nominal frequency is removed via the mixers as described above. The Doppler shift for the ASM viewed target is approximately defined as this difference between the received frequency and the nominal frequency. The frequency of the digital samples is

$$f_D = f_R - f_0 = \delta f + f_T \frac{v}{c}\cos\varphi + f_T \frac{v_T}{c}\cos\theta_T \qquad (2.149)$$

By examining some special cases, an intuitive understanding of this Doppler measurement can be developed using Figure 2.28. Consider the case of a target at rest ($v_T = 0$). For this case

$$f_D = \delta f + f_T \frac{v}{c}\cos\varphi \qquad (2.150)$$

If $\delta f = 0$ and the antenna is aligned with the ASM velocity vector ($\gamma = 0$), then the Doppler measurement is a measure of the target angle off bore sight (ψ)

$$f_D = f_T \frac{v}{c}\cos\psi \qquad (2.151)$$

This result is indicated in Figure 2.30. The range Doppler array resembles an image of the ocean scene being an array of measurements of echoes in range and angle relative to the ASM velocity vector.

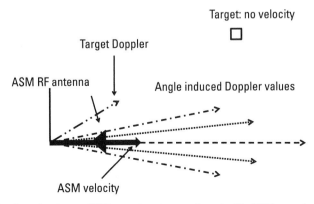

Figure 2.30 Doppler due to ASM speed, antenna aligned with ASM speed.

For the more general case shown in Figure 2.31, the Doppler is related to the target angle φ via

$$f_D = f_T \frac{v}{c} \cos\varphi \qquad (2.152)$$

Again the ASM parameters may be set in a variety of ways. The intent is to measure information about the target. It is convenient to set the transmit frequency to correct for Doppler-induced frequency because of ASM motion based on the look direction of the ASM antenna. The ASM processor knows the ASM platform speed and the look direction of the antenna relative to the ASM centerline. Set the transmit frequency as follows:

$$f_T = f_0 + \delta f \qquad (2.153)$$

$$\delta f = -f_0 \frac{v}{c} \cos\gamma \qquad (2.154)$$

$$|\delta f| = \left| f_0 \frac{v}{c} \cos\gamma \right| \ll f_0 \qquad (2.155)$$

$$f_D = -f_0 \frac{v}{c} \cos\gamma + f_0 \frac{v}{c} \cos\varphi \qquad (2.156)$$

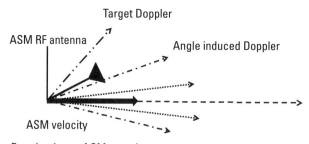

Figure 2.31 Doppler due to ASM speed.

Now any target at rest and on bore sight of the antenna ($\varphi = \gamma$) has zero Doppler frequency, as illustrated in Figure 2.32. After some simple algebra and approximations, (2.156) is approximately

$$f_D = -f_0 \frac{v}{c}\cos\gamma \cdot (1-\cos\psi) - f_0 \frac{v}{c}\sin\gamma \cdot \sin\psi \qquad (2.157)$$

$$f_D \approx -f_0 \frac{v}{c}\cos\gamma \cdot \frac{\psi^2}{2} - f_0 \frac{v}{c}\sin\gamma \cdot \psi \qquad (2.158)$$

$$f_D \approx -f_0 \frac{v}{c}\sin\gamma \cdot \psi \qquad \gamma \neq 0 \qquad (2.159)$$

$$f_D \approx -f_0 \frac{v}{c} \cdot \frac{\psi^2}{2} \qquad \gamma = 0 \qquad (2.160)$$

The antenna steering angle relative to the ASM centerline is γ. When the target is in view, the target angle off bore sight (ψ) is small. If the antenna scans (e.g., when in search mode $\gamma = [-45°$ to $+45°]$ or $[-90°$ to $+90°]$), the target Doppler will vary throughout the range Doppler array as a result of this antenna motion. Neglecting any target motion along the direction to the ASM, the Doppler measurement has a component that is a measure of the target angle relative to antenna bore sight (ψ).

The general expression for the corrected Doppler frequency measurement includes a term that measures the target speed along the direction to the ASM plus this antenna pointing angle induced term, as shown in Figure 2.33

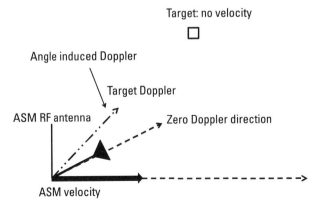

Figure 2.32 Corrected Doppler due to ASM speed.

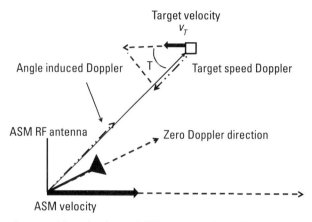

Figure 2.33 Corrected Doppler due to ASM speed, angle, and target velocity.

$$f_D = +f_0 \frac{v_T}{c}\cos\theta_T - f_0 \frac{v}{c}\cos\gamma \cdot (1-\cos\psi) - f_0 \frac{v}{c}\sin\gamma \cdot \sin\psi \qquad (2.161)$$

The case where γ is small will be considered at length in Chapter 6. In this case

$$f_D \approx +f_0 \frac{v_T}{c}\cos\theta_T - f_0 \frac{v}{c}\sin\gamma \cdot \sin\psi \qquad (2.162)$$

It is important to remember that the range Doppler measurement of a target by the ASM is typically a measure of the radial motion of the target combined with a measure of its angle off the ASM antenna bore sight. This will be examined in detail below. For now, the data may be considered to be associated with the geometry shown in Figure 2.34.

Now examine some particular issues related to the timing of the ASM pulsed Doppler radar sensor and its measurements. As indicated above the radar processor is collecting digital data for later processing. This processed data is required for selecting the target, extracting guidance information, and countering attempts by the fleet to corrupt or deny information via EA techniques.

Before an array of data can be collected the sensor hardware parameters must be determined and set. In Figure 2.35, this is represented by the arrows in the top row labeled by a nominal CPI number. In the figure, it is assumed that the system selects and sets hardware and software parameters defining CPI #0. At the appropriate time, the RF processing for CPI #0 begins and the CPI of data can begin to be collected by the RF processing. During this time interval, range compression may also be performed as data is positioned in a pulse number versus range array via the receivers and their ADCs. During this

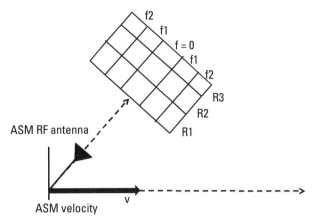

Figure 2.34 Symbolic data array for single CPI.

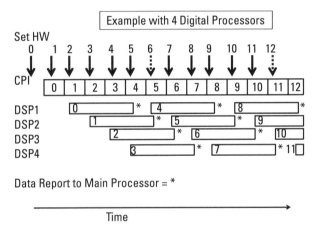

Figure 2.35 Processor timing illustration.

CPI the main processor in the ASM can determine the parameters for the next CPI and send messages to the appropriate hardware setting these parameters.

At this point, an array as shown on the left of Figure 2.29 has been generated for each receiver. After the entire CPI of data is collected and range compressed, and while the next CPI of data is being collected, this data array is *corner turned* in memory and begins Doppler processing. Data across the entire CPI for each range sample can be transferred to a digital signal processor (designated DSP number in Figure 2.35), while Doppler processing is conducted. This must be completed during the next CPI as shown in the figure.

This array (right side of Figure 2.28) is now (some time during CPI #1) available at this particular processor (DSP1). Various processing including extreme value measurements, data sample averages, noise estimation, and saturation events can be collected for AGC adjustment inputs. In addition, constant false alarm rate detection (CFAR), side lobe detections, and possible target detections can be performed for target parameter measurements, EP, target classifications, and monopulse processing. Since this processing can be extensive, the sensor typically has multiple digital processors available. As an example, the figure illustrates a case with four digital processors. The information from the digital processing is packaged and must be transferred to a main processor. It is the main processor that uses this information for changing RF parameters, target processing, guidance control, and AGC adjustments.

Typical timing is shown in the figure. As mentioned, DSP processing of these large data arrays may be extensive and time-consuming. There must be enough processors that all of this DSP processing can be completed and the reports sent to the main processor before the particular processor is needed for a subsequent CPI. In the illustration, the information collected during CPI #0 can be at the earliest impact parameter settings for CPI #6. Thus, while data collection may continue sequentially, the engineer should be aware that there is a significant (multiple CPIs) time delay between the measurements and the impact of the data measurements on the ASM. This delay is related to the architecture of the digital processing (e.g., the number of digital processors) of the seeker sensor.

Note in addition that, if the main processor decides to alter parameters in a significant manner (e.g., to change from search mode to track mode or to significantly change the characteristics of the CPI such as changing the waveform), there will be an interval of no changes to the information available until the processing pipeline can be refilled. In the case shown in Figure 2.35, changes made in CPI #6 will not be available until CPI #12. For example, if the ASM main processor decides to change the CPI characteristics at CPI #6, in this example there will be no useful measurement updates available until CPI #12 and the measurements from CPI #7 through CPI #11 are generally discarded.

References

[1] Stimson, G. W., *Introduction to Airborne Radar*, El Segundo, CA: Hughes Aircraft Company, 1983.

[2] Sullivan, R. J., *Microwave Radar Imaging and Advanced Concepts*, Norwood, MA: Artech House, 2000.

[3] Wiley, R. G., *Electronic Intelligence: The Analysis of Radar Signals*, Norwood, MA: Artech House, 1993.

[4] Tsui, J., *Digital Techniques for Wideband Receivers*, Norwood, MA: Artech House, 2001.

[5] Wehner, D., *High-Resolution Radar*, Boston, MA: Artech House, 1995.

[6] Schleher, D. C., *Electronic Warfare in the Information Age*, Norwood, MA: Artech House, 1999.

[7] Pace, P. E., *Detecting and Classifying Low Probability of Intercept Radar*, Norwood, MA: Artech House, 2009.

[8] MacGrath, D., "Analysis of Radome Induced Cross Polarization (U)," WL-TM-92-700-APN, USAF, Washington, DC, March 1992.

[9] Chen, V. C., *The Micro-Doppler Effect in Radar*, Norwood, MA: Artech House, 2011.

3
LPI Radar and EA Model

In this chapter, the information from the previous chapters is utilized to develop a physics-based mathematical model of the sensor of an ASM attacking a naval ship. This model is used in the later chapters to detail the variety of EP algorithms employed by the modern ASM to counter conventional EA. For simplicity, the model assumes that the ASM employs a sea-skimming flight profile, and the primary seeker sensor is an azimuth monopulse pulsed Doppler radar. The target ships employ a variety of EA assets including chaff, onboard digital RF memory (DRFM)-based repeaters, and noise or cover jamming, as well as various EA assets from off-board platforms and decoys. The goals of this chapter are to elaborate the mathematical model and to develop an engineering intuition for modern EW signal processing.

In the first section, the ASM model characteristics and specific signal processing parameters are discussed. The discussion will show the EW engineer some of the advantages of the LPI radar seeker. In particular, this section will improve the engineer's understanding of the advantages inherent in radar pulse compression and Doppler processing. A typical flight profile is described, and typical modes of the seeker sensor are discussed. The logical sequence of the ASM seeker tasks of target detection, identification, and guidance are explained.

In the second section, observations are made regarding the radar range equation. The concept of burn through is described for EA cover noise jamming.

It is clear that the source of the noise jamming late in the engagement should not be associated with the line of sight to any ship since the ship may eventually be detectable. It is shown how to mimic EA false targets with the proper radar cross-section (RCS) level.

Presently the array of range Doppler data is, at best, a crude image of the sea surface and targets. It is most useful to consider it as an array of digital data. In the third section, comments are made about the coming evolution of the ASM seeker sensor to the capability that image processing concepts may be profitably implemented. The reader needs to be aware of this near-term eventuality.

In the next section, the basic sensor model equations are developed for the echo from a simple point element of the ship target. These detailed mathematical expressions fully represent the signal portion of a single pixel of the range Doppler data array. The detailed expressions are simplified and modified in the later chapters to gain understanding of a variety of special cases of EW signal processing. For example, in Chapter 4, the expressions are modified to describe the relevant EP algorithms related to the true nature of the ship as an extended target.

In the fifth section of this chapter, the detailed sensor model equations are developed for a DRFM-based EA system. The DRFM-based device has the capability of recording an intercepted pulse from the ASM radar, modifying the pulse, and retransmitting the pulse back to the radar for the purpose of deception or seduction EA. The equations for the DRFM-based EA model will be used in the subsequent chapters to illustrate ASM EP algorithms.

An alternative EA mentioned previously is to generate noise in a major portion of the range Doppler data array for the purpose of covering the ship echo. The goal of this technique is to raise the general signal level in all pixels to mask the ship echo and thus deny the ASM seeker from measuring the target guidance parameters. A mathematical model for noise jamming is detailed and discussed.

In the final section, there is a summary discussion of the variety of model parameters that are commonly utilized for target classification and EP. Examples of specific sensor signal processing parameters are described. These parameters are discussed in detail in the later chapters and are used to quantify examples of EW signal processing. Finally, details of some typical EA strategies associated with the previous model examples are discussed. The final section contains a summary of the general model parameters and the key model expressions to be used later in this book.

3.1 ASM Model

In this first section, the ASM model characteristics and specific radar and signal processing parameters are discussed. The goals are to detail typical parameters for later use and to illustrate some of the inherent advantages as a seeker sensor of the coherent LPI radar over the classical noncoherent radar.

The ASM may be launched from land, from a surface ship, from a submerged submarine, or from an aircraft. The ASM can employ a variety of flight profiles ranging from sea skimming to a high-angle dive from a low-space trajectory. The ASM can be launched from close to the target ships or from a considerable distance from the fleet. A modern ASM can fly several hundreds of kilometers and use navigation systems to make course corrections at multiple check points.

For the basic ASM model herein, it is assumed that the operator has provided initial estimates of the target location and the target type. (If the initial target location results from acoustic sensors the target location may be ambiguous to within one of the several convergence zones.) The model developed herein assumes the ASM is initially flying below the radar horizon at a speed of about Mach 1 (about 300 m/s). At a range to the target of about 50 to 70 km, the ASM climbs to just above the radar horizon and makes a wide antenna sweep to verify the target parameters. This is a search mode or a reacquisition mode. This operation may take several seconds to ten seconds. This data refines and updates the target location and provides update data about target identification features. The ASM then dives back below the radar horizon and continues to approach the target. Figure 3.1 illustrates the model profile.

At a range to the target of about 20 km, the ASM begins its terminal engagement phase. This phase is the primary focus of this work. The ASM again climbs above the radar horizon and executes a search or reacquisition mode. The purpose is to reacquire or detect the target. This mode is accomplished by sweeping the antenna in azimuth and processing a swath of range Doppler data where the target is known to be from its prior information. Of course, the target may have moved up to 2 km (at 30 kts) from the last measured position. The antenna beamwidth of about 9° is about 3 km wide at this range. If the radar sweeps up to 45° to each side, the data may cover a patch of ocean surface about 40 km wide. The range swath may be as small as 4 km or as deep as 10 km or more.

Once the data is collected and analyzed, the ASM selects the most likely target. This is the seeker target classification mode. At this stage, the ASM may continue to process data from this single target, or it may continue to monitor

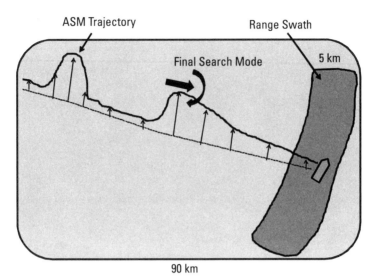

Figure 3.1 Model ASM flight profile.

many possible targets. Typically the antenna is pointed at the target of interest to maximize SNR. For the remainder of the engagement the ASM seeker is in the target track mode. The purpose of the track mode is to continuously collect accurate target measurements for the guidance subsystem. Table 3.1 summarizes the ASM trajectory and seeker modes for the model configuration.

During any of these modes, various EP algorithms are utilized to counter any EA from the fleet. For example, if the target is not detected during the search mode, but it is evident that the fleet is employing noise jamming, the ASM seeker may initiate a home on jam (HOJ) mode. In this mode the antenna points at the jamming source and flies toward the source assuming burn through may occur.

Table 3.1
Final Trajectory Time Line Summary

Time (sec)	Time to Go (sec)	Range (km)	Mode	Notes
0	200	60	Search	Pop up
130	70	21	Search	Pop up
140	60	18	Classify	EP
150	50	15	Track	Guide
200	0	0		Impact

Figure 3.2 illustrates a simple representation of the ASM subsystems of interest in this section. The transmitter, receivers, and antenna comprise the RF portion of the radar. The digital signal processing of the radar data arrays is accomplished in the digital processors represented in the figure as DSPs. The general processor is where the results of the signal processing are analyzed for mission objectives. The target detection, classification, and tracking decisions are made by this processor. In addition, commands are sent to the various subsystems for hardware and software functions, and measurements are sent to the guidance subsystem for ASM trajectory corrections.

The first task is to compare the classical noncoherent radar with the modern coherent radar for the seeker sensor application. The number of variables makes it difficult to compare the sensors. Care must be taken to relate as many of the common features to isolate the coherent features of the sensor. Consider first the search mode. In the typical sensor design, the false alarm rate (FAR) is initially set and then the detection performance is analyzed. The average time between false alarms is the inverse of the false alarm rate (FAR).

$$T_{\text{avg}} = \frac{1}{\text{FAR}} \qquad (3.1)$$

The probability of false alarm in a single cell is very small and the number of opportunities for a false alarm in this time is correspondingly large. Let n correspond to the number of opportunities for a false alarm in T_{avg}. The probability of one or more false alarms occurring in this time is

$$\text{Prob of 1 or more FAs in } T_{\text{avg}} = 1 - (1 - P_{\text{FA}})^n \qquad (3.2)$$

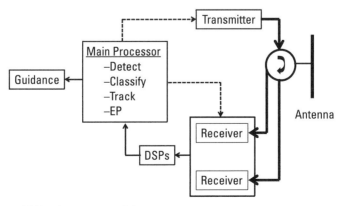

Figure 3.2 ASM subsystems model.

$$\text{Prob of 1 or more FAs in } T_{\text{avg}} \approx nP_{\text{FA}} - \frac{n(n-1)}{2}P_{\text{FA}}^2 + \ldots \quad (3.3)$$

$$\text{Prob of 1 or more FAs in } T_{\text{avg}} \approx nP_{\text{FA}} \quad (3.4)$$

If it is assumed that at least 1 false alarm occurs in this time

$$\text{Prob of 1 or more FAs in } T_{\text{avg}} \approx 1 \quad (3.5)$$

$$P_{\text{FA}} \approx \frac{1}{n} \quad (3.6)$$

With this crude, but useful approximation, the FAR is readily related to the number of false alarm opportunities that occur in this time defined in (3.1). (Throughout this section, the classical, noncoherent sensor is designated by subscript I and the modern, coherent sensor by subscript C.) The search mode is customarily divided into individual dwells on successive and particular space areas or volumes. Assume that the dwell for classical radar is a single pulse. Then the time for the dwell duration is the pulse repetition interval (PRI) (the time between pulses). The number of opportunities in the single dwell is equal to the number of range cells in the sampled range swath as shown in Figure 3.3.

Thus, the number of false alarm opportunities in the average time between false alarms can be calculated with respect to the number of range cells in the dwell (N_{RG}) as

$$N_I = \frac{N_{\text{RG}}}{\text{FAR} \cdot \text{PRI}} \quad (3.7)$$

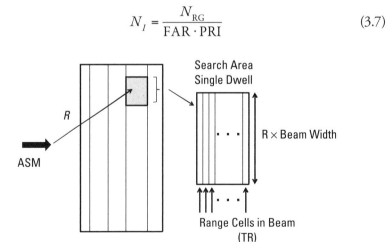

Figure 3.3 ASM search area for a single dwell.

Assume that the coherent radar has the same antenna beamwidth and dwells on the same range swath. However, the processing interval this time is a full CPI consisting of P pulses. The range returns from the P pulses are processed as a number of Doppler cells (N_D) and

$$N_C = \frac{N_{RG} \cdot N_D}{FAR \cdot CPI} \quad (3.8)$$

It is noted that the PRI can be related to the pulse repetition frequency (PRF) and that the following relationships relate the PRF to the maximum unambiguous range and the maximum Doppler velocity

$$R_{max} = \frac{c \cdot PRI}{2} \quad (3.9)$$

$$v_{max} = \frac{c \cdot PRF}{4 \cdot f_c} \quad (3.10)$$

$$PRF = \frac{1}{PRI} \quad (3.11)$$

If the PRF is greater than 2 kHz for the sensor with carrier frequency (f_c) equal to about 9.3 GHz, then there are Doppler cells corresponding to radial speeds in excess of 30 kts. For most naval ships, the maximum radial speed is about 30 kts. Higher speeds would not represent physically realizable targets. Assuming a low PRF so that there are no cells that relate to nonphysical targets, the following relationships lead to

$$CPI = P \cdot PRI \quad (3.12)$$

$$N_D = P \quad (3.13)$$

$$N_I = N_C \quad (3.14)$$

Thus, assuming these two sensors (one coherent and one noncoherent) have the same PRF, viewing the same size range swath, and with the same range resolution and antenna beamwidth, they meet the same FAR criterion with the same probability of a false alarm (PFA) for a sample cell. For the noncoherent sensor, the data sample is a single range sample for a single pulse. For the coherent sensor, the data sample is a single range Doppler measurement for a single CPI.

The next task is to analyze the detection performance of the two sensors. As stated above the detection performance is related to the SNR for a cell containing a target return. Following a standard radar reference [1–6], the radar range equation is recalled. Assume that the ASM radar transmits peak power in a single pulse P_{Pk} via a focused beam (G) in the direction of a target at range R. The power density at the target is

$$\text{Single Pulse Power at target} = \frac{P_{Pk} G(\psi)}{4\pi R^2} \qquad (3.15)$$

Defining the RCS of the target as the fraction of power transmitted back to the ASM antenna, the power density at the ASM antenna is

$$\text{Single Pulse Echo Power} = \frac{P_{Pk} G(\psi) \sigma_T}{(4\pi)^2 R^4} \qquad (3.16)$$

Relating the intercepting area of the antenna to its beam pattern and including a term for any losses gives the peak power into the receiver

$$\text{Power received} = \frac{P_{Pk} [G(\psi)]^2 \sigma_T \lambda^2}{(4\pi)^3 R^4 L} \qquad (3.17)$$

Several results from standard radar signal processing are recalled [5]. The maximum SNR is equal to the total energy in the received signal divided by the receiver noise. The total energy is the power times the pulse duration, and the output duration of the optimum matched filter is twice the pulse duration.

The power in the single pulse can be related to the average power using the PW and the PRI

$$P_{avg} = \frac{P_{Pk} \cdot PW}{PRI} = \frac{P_{Tot}}{PRI} \qquad (3.18)$$

As above, assume that the noncoherent radar uses a single pulse. Define its coherent processing time as the PRI. Similarly, the coherent processing time of the coherent radar includes the full set of P pulses and so

$$T_{coh_I} = PRI \qquad (3.19)$$

$$T_{coh_C} = PRI \cdot P = CPI \qquad (3.20)$$

The receiver noise terms are represented by the noise figure (F) and the temperature energy term ($kT_0 \times$ bandwidth [BW]), where BW is the receiver band width or the inverse of the matched filter duration. Remembering that the coherent system SNR increases with the number of pulses, the SNR for both systems can be written as

$$\text{SNR} = \frac{P_{\text{avg}} T_{\text{coh}} [G(\psi)]^2 \sigma_T \lambda^2}{(4\pi)^3 R^4 \cdot FL \cdot kT_0} \tag{3.21}$$

For convenience, this expression is written (for each radar) as

$$\text{SNR}_I = \frac{[G(\psi)]^2 \lambda^2 \text{PRI}}{(4\pi)^3 \cdot FL \cdot kT_0} \cdot \frac{P_{\text{avg}I} \sigma_I}{R_I^4} \tag{3.22}$$

$$\text{SNR}_C = \frac{[G(\psi)]^2 \lambda^2 \text{PRI}}{(4\pi)^3 \cdot FL \cdot kT_0} \cdot \frac{P_{\text{avg}C} P\sigma_C}{R_C^4} \tag{3.23}$$

Assuming that both systems have the same FAR, the same noise characteristics, and the same PRF, the detection performance of each is equal if

$$\frac{P_{\text{avg}C} P\sigma_C}{R_C^4} = \frac{P_{\text{avg}I} \sigma_I}{R_I^4} \tag{3.24}$$

Consider first the case in which both sensors have the same average power. Since it was assumed that the PRF is the same for both the systems, (3.18) gives

$$P_{\text{Pk}C} \cdot \text{PW}_C = P_{\text{Pk}I} \cdot \text{PW}_I \tag{3.25}$$

There are several reasons to not set these individual parameters equal. As stated previously, the noncoherent system requires a small PW to achieve a useful range resolution. The coherent system can use various waveform modulations and pulse compression techniques to achieve a fine range resolution. Thus, the coherent system peak power can be much lower than the noncoherent system peak power. This lower peak power makes transmitter components and receiver components more efficient. In addition, a low peak power with a modulated and wide pulse makes it much more difficult for the EW support systems in the fleet to detect and intercept the ASM coherent (LPI) radar transmissions.

For example, assume that the noncoherent system has a PW of 0.125 μs. Coherent radar can achieve the same range resolution using a modulated pulse with a BW of 8 MHz. Postulating the PW of the coherent system to be 8 μs and the peak power to be 600W, the noncoherent system must have a peak power of 38.4 kW or 18-dB higher power.

From (3.24) for the case of equal average power, the detection performance is the same if

$$\frac{P\sigma_C}{R_C^4} = \frac{\sigma_I}{R_I^4} \qquad (3.26)$$

As an example, consider the case in which 16 pulses (P) are coherently combined via Doppler processing. Assume also that the noncoherent system can detect a 40-dBsm target at a 20-km range. For this example the coherent sensor can detect the same size target (40 dBsm) at twice the range or 40 km. Or, equivalently, from (3.26) the coherent sensor can detect a much smaller target (28 dBsm) at this same range (20 km).

The analysis to this point has assumed that the noise of interest is receiver white noise. White noise distributes evenly in the several Doppler filters. Sea clutter presents another and very significant interference as colored noise. The standard model of sea clutter is that the main lobe clutter is concentrated at zero radial speed, and the side lobe clutter spreads among the nonzero Doppler frequencies. Also the side lobe clutter is typically about 30 dB lower than the peak clutter as shown in Figure 3.4.

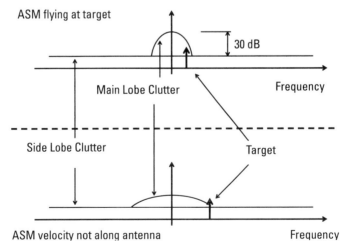

Figure 3.4 Clutter Doppler characteristics.

Figure 3.5 illustrates the geometry. From Chapter 2, (2.162) as applied to a target echo gives an approximation for the Doppler frequency as

$$f_D \approx +f_0 \frac{2v_T}{c}\cos\theta_T - f_0 \frac{2v}{c}\sin\gamma \cdot \sin\psi \quad (3.27)$$

The top portion of Figure 3.4 qualitatively illustrates the spectrum for a target with low radial speed and an ASM flying directly at the target and with its antenna pointing at the target. If the geometry is more general as illustrated in Figure 3.5, then the main lobe clutter smears and becomes lower. Also, if the target is not directly in the center of the beam, there is an added contribution to its measured Doppler frequency. As a result the spectrum is more like the lower portion of Figure 3.4. In this more general case, the target is more apt to be detected. The probability of a false alarm in the dwell is equal to the inverse of the number of opportunities in the dwell, assuming each opportunity is equally likely.

A standard model of the sea clutter is to consider the clutter an area-wide target. At low grazing angle, the clutter is represented by a target of –35 dB/area. Using an 8° beamwidth and a range resolution of 20m, the area is 47 dB and the main lobe clutter is 12 dB. At a speed of Mach 1 and a look angle of 5°, the main lobe clutter spreads over several Doppler cells and is comparable to a small target of about 20 dB in these low Doppler cells. Thus, at shorter ranges the detection of targets in low Doppler cells may be clutter limited rather than receiver noise limited.

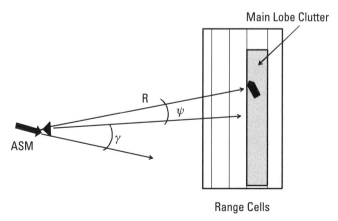

Figure 3.5 Clutter and target geometry.

Using a simple clutter model the clutter-to-noise ratio (CNR) is

$$\text{CNR}_I = \frac{[G(\psi)]^2 \lambda^2 \text{PRI}}{(4\pi)^3 \cdot FL \cdot kT_0} \cdot \frac{P_{\text{avg}I} \sigma_0 A_I}{R_I^4} \tag{3.28}$$

$$\text{CNR}_C = \frac{[G(\psi)]^2 \lambda^2 \text{PRI}}{(4\pi)^3 \cdot FL \cdot kT_0} \cdot \frac{P_{\text{avg}C} P \sigma_0 A_C}{R_C^4} \tag{3.29}$$

Combining these expressions with (3.22) and (3.23) the signal-to-clutter ratio (SCR) is

$$\text{SCR}_I = \frac{\sigma_T}{\sigma_0 \cdot A_I} \tag{3.30}$$

$$\text{SCR}_C = \frac{\sigma_T}{\sigma_0 \cdot A_C} \tag{3.31}$$

If the areas of the antenna footprint on the sea surface are the same, the clutter limited detection performance for targets in low Doppler cells is the same for the noncoherent system and the coherent system. Obviously the existence of clutter spikes is a special case. Clutter spikes can have a significant cross section and a significant nonzero Doppler speed. These spikes can appear similar to low-level targets. Table 3.2 presents a summary of three examples of these results.

Thus far the parameters have been kept as similar as possible. Table 3.3 lists a more realistic set of parameters for noncoherent ASM radar and for coherent ASM radar. In this table, it is assumed that the target amplitude is Rayleigh distributed and the area of the cell (for clutter calculation) is

$$A = \delta R \cdot \theta \cdot R \tag{3.32}$$

The improvement of the coherent example in the noise limited region results from the assumption of the interpulse coherent gain (CPI versus PRI). This improvement is somewhat mitigated by the much lower average power of 9.4 dB. In the clutter limited region, there is a 10-dB improvement resulting from the intrapulse coherent gain. Figures 3.6 and 3.7 illustrate the effect. While the signal (target) is completely within the cell in both the cases, the clutter area is much compressed for the coherent sensor.

Table 3.2
Comparable ASM Radar Systems Summary

Parameter	Noncoherent	Coherent 1	Coherent 2
FAR	1/hr	1/hr	1/hr
PFA/cell	4.6×10^{-10}	4.6×10^{-10}	4.6×10^{-10}
Prob det	0.5	0.5	0.5
SNR (reqd)	14.8 dB	14.8 dB	14.8 dB
Range swath (km)	6	6	6
λ (cm)	3	3	3
P_{Pk}	1 kW	1 kW	15.6 W
PRF (kHz)	2	2	2
PW (μs)	0.125	0.125	8
P_{avg} (W)	0.25	0.25	0.25
G (dB)	25	25	25
Beamwidth (deg)	8	8	8
Bandwidth (MHz)	8	8	8
δR (m)	20	20	20
Pulses per CPI	1	16	16
F (dB)	3	3	3
L (dB)	10	10	10
σ_T (dBsm)	35	35	35
Detection R (km)	9.4	18.7	18.7

Table 3.3
Systems Summary

Parameter	Noncoherent	Coherent
FAR	1/hr	1/hr
PFA/cell	3.5×10^{-8}	3.5×10^{-10}
Prob det	0.5	0.5
SNR (reqd)	14.2 dB	14.1 dB
Range swath (km)	3	3
λ (cm)	3	3

Table 3.3 (continued)

Parameter	Noncoherent	Coherent
P_{Pk}	35 kW	200W
PRF (kHz)	2	2
PW (μs)	1.0	20.0
P_{avg} (W)	70	8
G (dB)	25	25
Beamwidth (deg)	8	8
Bandwidth (MHz)	1	10
δR (m)	150	15
Pulses per CPI	1	16
F (dB)	3	3
L (dB)	10	10
σ_{TT} (dBsm)	35	35
Detection R (km)	39.6	46.2
SCR (at 20 km)	13.8 dB	23.8 dB
SCR (at 30 km)	12.0 dB	22.0 dB

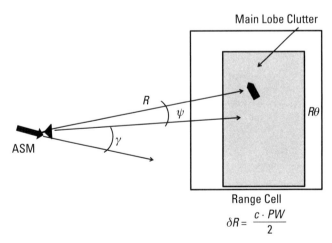

Figure 3.6 Noncoherent sensor range cell area.

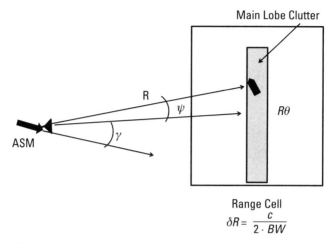

Figure 3.7 Coherent sensor range cell area.

Thus far the clutter area has been represented by (3.32). For the high-speed maneuvering ASM, there is a beam sharpening effect on the cell area. In this case the area is represented as (3.33) and illustrated in Figure 3.8.

$$A = \delta R \cdot \delta \psi \cdot R \qquad (3.33)$$

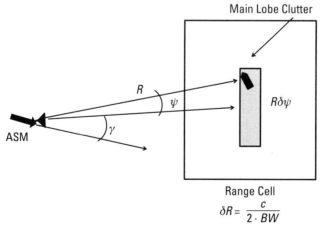

Figure 3.8 Beam sharpening effect on clutter.

Finally, it is noted that the tracking performance (as with most radar measurements) improves with SNR. The SNR increases approximately as the range to the fourth power.

$$\text{Measurement Variance} = \frac{k}{\text{SNR}} \quad (3.34)$$

3.2 Radar Range Equations and Burn Through

Returning to the model information above, examine more closely the echo signal and EA for the case of cover noise jamming. It was noted that the power density at the target is

$$\text{Single Pulse Power at target} = \frac{P_{Pk} G(\psi)}{4\pi R^2} \quad (3.35)$$

Defining the target RCS as the fraction of power transmitted back to the ASM antenna, the power density at the ASM antenna is

$$\text{Single Pulse Echo Power} = \frac{P_{Pk} G(\psi) \sigma_T}{(4\pi)^2 R^4} \quad (3.36)$$

Relating the intercepting area of the antenna to its beam pattern and neglecting the losses gives the peak power for a single pulse into the receiver

$$\text{Power received} = \frac{P_{Pk} [G(\psi)]^2 \sigma_T \lambda^2}{(4\pi)^3 R^4} \quad (3.37)$$

For a cover noise jamming system the equivalent expression for (3.36) relates the power sent to the ASM radar as

$$\text{EA power} = \frac{P_J G_J(\psi')}{4\pi R^2} \quad (3.38)$$

The subscript J indicates the peak power and the antenna gain are the values for the EA system. The angle (ψ') indicates the angle from the EA system to the ASM relative to the EA system antenna. While the jamming signal is transmitted continuously, the EA power noise level received via the ASM antenna during the single pulse is

$$\text{EA power received} = \frac{P_J G(\psi) G_J(\psi') \lambda^2}{(4\pi)^2 R^2} \quad (3.39)$$

The fleet distribution is typically small relative to the ASM detection range. Assume for simplicity that the target and the EA are at approximately the same range. (It is noted that the signal level has pulse compression gain. This factor is generally mitigated to some extent by capturing the ASM transmit pulse and repeating a portion of the pulse to regain a portion of the pulse compression gain. This technique is limited since this may lead to clear regions in the range Doppler array. This aspect is ignored at this time.)

Recall that during a single CPI the target signal power increases as the number of pulses squared, while the cover noise EA power level increases like thermal noise or as the number of pulses. Thus, the magnitude squared ratio of target echo signal to that of the EA noise level is the signal-to-jamming ratio (SJR)

$$\text{Ratio target to EA power} = \text{SJR} = \frac{P_{\text{Pk}} \sigma_T P}{P_J} \cdot \frac{G(\psi)}{G_J(\psi')} \cdot \frac{1}{4\pi R^2} \quad (3.40)$$

If the EA is cover jamming to mask the target the goal is to make this ratio as small as possible, preferably much smaller than 1. The minimum of the first term is when the EA system emits its maximum power. The second term can be reduced if the ASM sensor points in the direction of the EA system and the EA system is not the direction of the target. That is, the goal is to have the high-value unit (HVU) and the EA system at different angles relative to the ASM and to have the ASM point at the EA and away from the HVU as early in the engagement as possible. As long as the ASM is in HOJ mode, this term will continue to decrease.

The last term is small but continually increasing as the ASM approaches the fleet. The variation of the inverse range squared means that the SJR increases by 6 dB whenever the range reduces by half. If the target angle and the EA angle are not that significantly different, this ratio increases until the target may be detected. It is said that the target signal *burns through* the cover noise jamming. This is conceptually illustrated in Figure 3.9.

For completeness, it is mentioned that instead of cover noise EA, the EA system may generate one or more false target echoes. From (3.35) the EA system receives a single pulse power level of

$$\text{Single Pulse Power EA received} = P_R = \frac{P_{\text{Pk}} G(\psi'') A_J}{4\pi R^2} \quad (3.41)$$

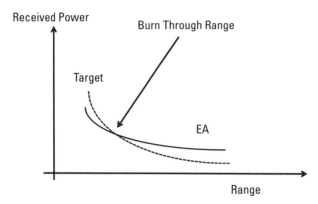

Figure 3.9 Burn through range of the target echo.

$$A_J = \frac{G_J(\psi')\lambda^2}{4\pi} \qquad (3.42)$$

Assuming that the EA system false target is generated with a fraction of the peak EA system power (indicated by the fraction r), the single pulse level of the false target at the ASM is from (3.39)

$$\text{EA Power received} = \frac{rP_J G(\psi'')G_J(\psi')\lambda^2}{(4\pi)^2 R^2} \qquad (3.43)$$

The ASM system would estimate the false target RCS as

$$\sigma_J = \frac{rP_J A_J}{P_R} \qquad r \leq 1 \qquad (3.44)$$

It is noted that all of these factors are known at the EA system if it successfully intercepts the ASM radar pulse. Thus, if this expression can be satisfied (without saturating the EA transmitter) a valid RCS level false target can be generated.

3.3 Range Doppler Map and Imaging

At each CPI the ASM sensor is presented with an array of data. The data is labeled as originating with a specific range measured via the time delay between the transmit pulse and the echo (δt) and Doppler (a function of the ASM antenna angle, the ASM velocity, and the target radial speed). The geometry is illustrated in Figure 3.10. The following expressions relate the several geometry measurements and their respective resolutions.

$$R = \frac{c \cdot \delta t}{2} \quad (3.45)$$

$$\delta R = \frac{c}{2 \cdot BW} \quad (3.46)$$

$$f_D \approx +f_0 \frac{2v_T}{c} \cos\theta_T - f_0 \frac{2v}{c} \sin\gamma \cdot \sin\psi \quad (3.47)$$

$$\delta v_T = \frac{c}{2 f_0 \cdot CPI} \quad (3.48)$$

The data is also represented in angle space as having originated from a particular angle relative to the antenna bore sight pointing direction via the monopulse technique. Figure 3.11 illustrates one of the receiver data arrays.

Utilizing typical parameters similar to those in [4], the ASM sensor is capable of generating an image in range/Doppler/angle space. However, the pixels or cells in the data array do not typically have a fine resolution. Typical ASM resolutions are approximately

$$\delta R = 20 \text{m} \quad (3.49)$$

$$\delta v_T = 4 kts = 2 \text{ m/sec} \quad (3.50)$$

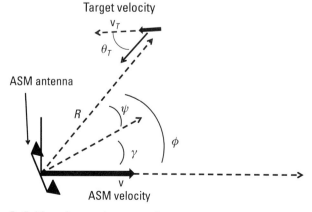

Figure 3.10 Definition of geometry parameters.

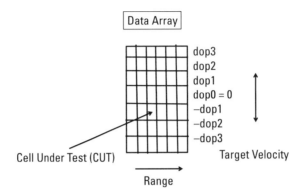

Figure 3.11 Illustration of one receiver data array.

$$\delta\psi = 0.1 \text{deg} \qquad (3.51)$$

Most standard naval targets reside mainly in a single range Doppler cell when viewed from the side. At any other aspect angle the targets usually occupy multiple cells. The only HVU that always resides in multiple cells is the aircraft carrier. Table 3.4 provides the general spatial dimensions of typical naval ships.

For now, it is noted that the Doppler measurement is a combination of target radial speed and angle off antenna bore sight. As ASM radar sensors improve, it is expected that the display will be closer to actual imaging such as synthetic aperture radar (SAR) or inverse SAR (ISAR) imaging. At the present time the image is limited at best. In the future, sophisticated image processing techniques are expected to be implemented in ASM sensor processing. However, the present resolutions are sufficient for many sophisticated signal processing techniques. These are discussed below.

Table 3.4
Spatial Dimensions of Naval Targets

Surface Ship Class	Length (m)	Beam (m)
Destroyer	14	135
Cruiser	17	175
Aircraft carrier	78	335

3.4 Target Scatter Model

Following [1–4] and repeating the analysis from above the radar range equation is recalled. Assume the ASM radar transmits a sequence of pulses with peak power P_{Pk} via a focused Σ beam (G_Σ) in the direction of a target at range R. The power density at the target is

$$\text{Power density at target} = \frac{P_{Pk} G_\Sigma(\psi)}{4\pi R^2} \quad (3.52)$$

Defining the target RCS as the fraction of power transmitted back to the ASM antenna, the power density at the ASM antenna is

$$\text{Power density at antenna} = \frac{P_{Pk} G_\Sigma(\psi) \sigma_T}{(4\pi)^2 R^4} \quad (3.53)$$

Relating the intercepting area of the antenna to its beam pattern gives the peak power into the Σ receiver for a single pulse

$$\text{Power received} = \frac{P_{Pk} [G_\Sigma(\psi)]^2 \sigma_T \lambda^2}{(4\pi)^3 R^4} \quad (3.54)$$

The amplitude squared of the return for a coherent sum of P pulses in a single CPI is

$$\text{Power received } P \text{ pulses} = \frac{P_{Pk} [G_\Sigma(\psi)]^2 \sigma_T \lambda^2 P^2}{(4\pi)^3 R^4} \quad (3.55)$$

For completeness the SNR of this return as indicated above is

$$\text{SNR} = \frac{P_{avg} T_{coh} [G_\Sigma(\psi)]^2 \sigma_T \lambda^2}{(4\pi)^3 R^4 \cdot FL \cdot kT_0} \quad (3.56)$$

The references give a thorough explanation of these approximations for the echo power. These results are based on the solution to Maxwell's equations at range R using spherical spreading of the E field and B field radiation. They are based on the result that the power density is proportional to the square of the E field and

$$\text{Power density at range } R \approx E^T \cdot E^* = |E|^2 \quad (3.57)$$

The primary interest of this work is in the EP techniques employed by the ASM radar sensor. This necessitates a closer look at the antenna and the coherent phase properties of the complex electric field of the received signal. A typical ASM antenna consists of a reflector with multiple feeds or a flat plate of a two-dimensional array of dipoles that are coherently combined. In each case the antenna can be viewed as comprised of four distinct quadrants. Each quadrant is a separate subarray antenna that can be treated in the manner of the linear array discussed above. Being interested only in the azimuth direction in this work, consider the ASM RF antenna as composed of two subarrays (the left half and the right half) designated U and L.

The ASM subarrays are characterized by their complex beam patterns that are components of the two-dimensional (in polarization space) vector. There have been many representations of the radar polarization, including the Jones vector and the location on the Poincare sphere [5]. In this work, the full antenna beam patterns are represented by more intuitive two-dimensional vectors as described.

The radar antenna polarization is represented by a two-dimensional vector notation borrowed from quantum physics. The transmit pattern is designated herein via a Ket vector. The Bra vector (receive antennas) is the complex conjugate transpose of the Ket vector by reciprocity. The assumption throughout is that these patterns are a result of the subantennas looking through the ASM radome [3]. The polarization component p is the polarization direction of the Σ beam pattern on bore sight and n is the orthogonal component. For example, p may refer to the vertical linear and n to the horizontal linear. The g functions are complex beam pattern functions of the azimuth angle ψ (not shown) and allow for general elliptical polarizations. Thus, the transmit antenna patterns for the U antenna subarray and for the L antenna subarray are the Ket vectors

$$|U\rangle = \begin{bmatrix} g_U^p \\ g_U^n \end{bmatrix} \tag{3.58}$$

$$|L\rangle = \begin{bmatrix} g_L^p \\ g_L^n \end{bmatrix} \tag{3.59}$$

The corresponding receive patterns at the same angle (not shown) off bore sight are the corresponding Bra vectors

$$\langle U| = \begin{bmatrix} g_U^{p*} & g_U^{n*} \end{bmatrix} \tag{3.60}$$

$$\langle L = \begin{bmatrix} g_L^{p*} & g_L^{n*} \end{bmatrix} \quad (3.61)$$

As before, assume the transmit signal consists of a carrier frequency f_T shifted from the desired receiver frequency f_0 to compensate for ASM platform motion-induced Doppler

$$f_T = f_0 + \delta f \quad (3.62)$$

(For simplicity, any phase modulation term for pulse compression is neglected in these expressions for now. The phase modulation can be added as required.) The signal at the object shown in Figure 3.12 is composed of the sum of the transmissions from the U and L subarrays.

Assuming the amplitude from the above discussion of the radar range equation, the resultant signals are slightly offset in phase relative to each other as a result of the slightly differing ranges to the individual subantennas

$$\text{Signal at target} = U\rangle \cdot \cos\left[2\pi f_T\left(t - \frac{R_U}{c}\right)\right] + L\rangle \\ \cdot \cos\left[2\pi f_T\left(t - \frac{R_L}{c}\right)\right] \quad (3.63)$$

Setting the antenna spacing to d, the expressions for the ranges from each subantenna are approximately

$$R_U = R_0 - vt\cos(\varphi) - v_T t\cos(\theta_T) - \Delta R \quad (3.64)$$

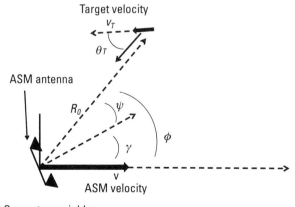

Figure 3.12 Geometry variables.

$$R_L = R_0 - vt\cos(\varphi) - v_T t \cos(\theta_T) + \Delta R \tag{3.65}$$

$$\Delta R = \frac{d}{2} \cdot \sin(\psi) \tag{3.66}$$

$$\Phi \equiv \frac{\Delta R}{\lambda} = \frac{d}{2\lambda} \cdot \sin(\psi) \tag{3.67}$$

Define the ASM antenna sum and azimuth delta transmit patterns as

$$\Sigma\rangle \equiv U\rangle + L\rangle \tag{3.68}$$

$$\Delta\rangle \equiv U\rangle - L\rangle \tag{3.69}$$

Since δf is much less than f_0, the transmit signal at the target is approximately

$$\arg 1 = f_0 t + \delta f t + \frac{vt\cos(\varphi)}{\lambda} + \frac{v_T t \cos(\theta_T)}{\lambda} - \frac{R_0}{\lambda} \tag{3.70}$$

$$\begin{aligned}\text{Signal at target} = {}& \Sigma\rangle \cos(2\pi\arg 1)\cos(2\pi\Phi) \\ & - \Delta\rangle \sin(2\pi\arg 1)\sin(2\pi\Phi)\end{aligned} \tag{3.71}$$

This signal reflects from a scatter element represented by the 2 × 2 (polarization) scatter matrix Ω (a generalization of the RCS given above). The echoes received at the two antenna subarrays (U and L) are combined in the antenna hybrid to form two received signals for the Σ and Δ receiver channels.

$$\langle\Sigma \equiv \langle U + \langle L \tag{3.72}$$

$$\langle\Delta \equiv \langle U - \langle L \tag{3.73}$$

Restoring the overall amplitude the target signals in the two receivers are

$$\arg 2 = f_0 t + \delta f t + \frac{2vt\cos(\varphi)}{\lambda} + \frac{2v_T t \cos(\theta_T)}{\lambda} - \frac{2R_0}{\lambda} \tag{3.74}$$

$$\begin{aligned}\Sigma = A_S(R)\big[& \cos(2\pi\arg 2)\cos^2(2\pi\Phi)\langle\Sigma|\Omega|\Sigma\rangle \\ & - \cos(2\pi\arg 2)\sin^2(2\pi\Phi)\langle\Delta|\Omega|\Delta\rangle \\ & - \sin(2\pi\arg 2)\cos(2\pi\Phi)\sin(2\pi\Phi)(\langle\Delta|\Omega|\Sigma\rangle + \langle\Sigma|\Omega|\Delta\rangle)\big]\end{aligned} \tag{3.75}$$

$$\Delta = \Big[A_S(R)[\cos(2\pi \arg 2)\cos^2(2\pi\Phi)\langle\Delta|\Omega|\Sigma\rangle$$
$$-\cos(2\pi \arg 2)\sin^2(2\pi\Phi)\langle\Sigma|\Omega|\Delta\rangle \qquad (3.76)$$
$$-\sin(2\pi \arg 2)\cos(2\pi\Phi)\sin(2\pi\Phi)\big(\langle\Sigma|\Omega|\Sigma\rangle + \langle\Delta|\Omega|\Delta\rangle\big) \Big]$$

Now introduce the following variables and make the appropriate approximations for the range sample for pulse p with PRI

$$\text{PRI} = T \qquad (3.77)$$

$$\delta f = \frac{-2v\cos\gamma}{\lambda} \qquad (3.78)$$

$$\beta = \frac{-4vT\sin\gamma}{\lambda} \qquad (3.79)$$

$$f_D = \frac{2v_T \cos\theta_T}{\lambda} \qquad (3.80)$$

The variable in (3.78) is the offset of the transmit frequency from the receiver nominal carrier frequency previously defined in (3.62). The variable in (3.79) is a dimensionless variable representing ASM platform speed orthogonal to antenna pointing. The variable in (3.80) represents the target radial speed portion of the Doppler frequency.

The target echo portion of the RF processing output at the ADC for pulse p is

$$\Sigma = A_S(R)e^{\left[2\pi i\left\{(\beta\Phi + f_D T)p - \frac{2R_0}{\lambda}\right\}\right]} \Big[\cos^2(2\pi\Phi)\langle\Sigma|\Omega|\Sigma\rangle$$
$$-\sin^2(2\pi\Phi)\langle\Delta|\Omega|\Delta\rangle \qquad (3.81)$$
$$+ i\cos(2\pi\Phi)\sin(2\pi\Phi)\big(\langle\Delta|\Omega|\Sigma\rangle + \langle\Sigma|\Omega|\Delta\rangle\big)\Big]$$

$$\Delta = A_S(R)e^{\left[2\pi i\left\{(\beta\Phi + f_D T)p - \frac{2R_0}{\lambda}\right\}\right]} \Big[\cos^2(2\pi\Phi)\langle\Delta|\Omega|\Sigma\rangle$$
$$-\sin^2(2\pi\Phi)\langle\Sigma|\Omega|\Delta\rangle \qquad (3.82)$$
$$+ i\cos(2\pi\Phi)\sin(2\pi\Phi)\big(\langle\Sigma|\Omega|\Sigma\rangle + \langle\Delta|\Omega|\Delta\rangle\big)\Big]$$

After collecting the outputs for P pulses of the CPI at this point in the processing, both sets of data are generally integrated via a discrete Fourier transform to translate the P time samples in each range cell into P Doppler samples for that range cell as discussed in Chapter 2. (A window function in time may be applied to lower the Doppler filter side lobes.) Thus, the samples in range cell at R_0 peak in the Σ channel at Doppler value

$$\text{Doppler frequency} \cdot T = \beta\Phi + f_D T \tag{3.83}$$

Combining this result with the amplitude result above the range Doppler cell terms of both receiver arrays at the Σ channel peak amplitude are

$$\Sigma = PA_S(R) \cdot e^{\left[-2\pi i \frac{2R_0}{\lambda}\right]} \Big[\cos^2(2\pi\Phi)\langle\Sigma|\Omega|\Sigma\rangle \\ - \sin^2(2\pi\Phi)\langle\Delta|\Omega|\Delta\rangle \\ + i\cos(2\pi\Phi)\sin(2\pi\Phi)\big(\langle\Delta|\Omega|\Sigma\rangle + \langle\Sigma|\Omega|\Delta\rangle\big) \Big] \tag{3.84}$$

$$\Delta = PA_S(R) \cdot e^{\left[-2\pi i \frac{2R_0}{\lambda}\right]} \Big[\cos^2(2\pi\Phi)\langle\Delta|\Omega|\Sigma\rangle \\ - \sin^2(2\pi\Phi)\langle\Sigma|\Omega|\Delta\rangle \\ + i\cos(2\pi\Phi)\sin(2\pi\Phi)\big(\langle\Sigma|\Omega|\Sigma\rangle + \langle\Delta|\Omega|\Delta\rangle\big) \Big] \tag{3.85}$$

The dominant terms are the terms quadratic in the Σ beam patterns and p polarization. Using σ_{pp} for the dominant component of Ω, these dominant terms are

$$\Sigma = PA_S(R) \cdot e^{\left[-2\pi i \frac{2R_0}{\lambda}\right]} \Big[\cos^2(2\pi\Phi)\big(|g_\Sigma^p(\psi)|\big)^2 \sigma_{pp} \Big] \tag{3.86}$$

$$\Delta = PA_S(R) \cdot e^{\left[-2\pi i \frac{2R_0}{\lambda}\right]} \Big[\cos^2(2\pi\Phi) g_\Delta^{p*}(\psi) \cdot g_\Sigma^p(\psi) \cdot \sigma_{pp} \\ + i\cos(2\pi\Phi)\sin(2\pi\Phi)\big(|g_\Sigma^p(\psi)|\big)^2 \sigma_{pp} \Big] \tag{3.87}$$

Classical ASM sensor processing, indicated in Figure 3.13, uses only Σ channel data for detection, target classification, and most EP efforts. The data is commonly arranged as an array in range and Doppler for the Σ channel and similarly for the Δ channel. The Σ channel data array is customarily processed

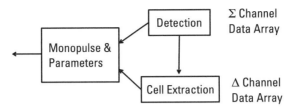

Figure 3.13 Typical ASM sensor processing.

via a CFAR type algorithm to detect cells containing significant amplitude data for potential target detection. These terms are then examined for target classification and EP. During track mode, various tracking gates are formed around this region in the array to protect from various EA such as seduction techniques. In addition, the corresponding cell of the Δ channel is used to form the monopulse ratio to extract additional guidance information.

From (3.86) and (3.87), the monopulse ratio is approximately

$$\frac{\Delta}{\Sigma} = i\tan(2\pi\Phi) + \frac{g_\Delta^{p*}(\psi)}{g_\Sigma^{p*}(\psi)} \quad (3.88)$$

From (3.67), this expression can be simplified to

$$\frac{\Delta}{\Sigma} = i\frac{d\pi}{\lambda}\cdot\psi \quad (3.89)$$

The monopulse concept is discussed further below.

3.5 Repeater EA Model and the DRFM

The DRFM device allows the EA system to store a high fidelity copy of the sensor transmitted signal and to then repeat it back with a variety of modifications [2, 6]. The DRFM-based EA system can return one or more false echoes with the intent of deception. Or the EA system can repeat back cover or noise jamming with the intent of masking the echo from the HVU and/or other targets.

Repeating the above analysis for a DRFM-based EA system, consider first the radar range equation. The ASM radar transmits peak power P_{PK} to the jamming system at range R, delivering peak power density

$$\text{Power density at EA} = \frac{P_{pk}G_\Sigma(\psi)}{4\pi R^2} \qquad (3.90)$$

The EA system receives this signal with power level

$$\text{EA received power} = \frac{P_{pk}G_\Sigma(\psi)G_J(\psi')\lambda^2}{(4\pi)^2 R^2} \qquad (3.91)$$

This power is received by the jamming receiver and the EA system then transmits a signal back to the ASM antenna. Using a J subscript to indicate EA system parameters the power density received at the ASM antenna is

$$\text{EA power density at ASM} = \frac{P_J G_J(\psi')}{4\pi R^2} \qquad (3.92)$$

Relating the intercepting area of the ASM sensor antenna to its beam pattern gives the power into the ASM Σ receiver

$$\text{Power received} = \frac{P_J G_\Sigma(\psi)G_J(\psi')\lambda^2}{(4\pi)^2 R^2} \qquad (3.93)$$

This approximation is the radar range equation for the single pulse of the false target or for cover jamming. Most notably the power is related to the inverse square of the range. For a viable false target the power for a full CPI varies as P^2. For cover noise EA the power for a full CPI varies as P. Various noise jamming techniques use a partial pulse to achieve additional gain.

Again the primary interest is in the electric field at the ASM radar sensor. Examining the phase of the received EA signal electric field, the signal at the EA system is the same as in the target expressions

$$\arg 1 = f_0 t + \delta ft + \frac{vt\cos(\varphi)}{\lambda} + \frac{v_T t\cos(\theta_T)}{\lambda} - \frac{R_0}{\lambda} \qquad (3.94)$$

$$\begin{aligned}\text{Signal at EA} = \Sigma\rangle\cos(2\pi\arg 1)\cos(2\pi\Phi) \\ -\Delta\rangle\sin(2\pi\arg 1)\sin(2\pi\Phi)\end{aligned} \qquad (3.95)$$

The phase variation of this signal is received by the EA system as

$$\text{EA received} = \langle \Sigma_J | \Sigma \rangle \cos(2\pi \arg 1)\cos(2\pi\Phi) \\ - \langle \Sigma_J | \Delta \rangle \sin(2\pi \arg 1)\sin(2\pi\Phi) \tag{3.96}$$

This signal is RF processed by the EA system and stored via the DRFM. It is transmitted back to the ASM with perhaps a delayed range and a pseudo Doppler phase shift. (For a long pulse the EA may even transmit a partial pulse.) For a false target the results are similar to the above and will be indicated below. For now, assume the signal is emitted back to the ASM sensor as cover noise. Assume a simplistic noise model such that a random phase η_p^J is added to the return for pulse p, perhaps varying with range. Following the logic as in the previous section leads to the following results for the cover noise EA received by the ASM receivers

$$\arg 2 = f_0 t + \delta f t + \frac{2vt\cos(\varphi)}{\lambda} - \frac{2R_J}{\lambda} + \eta_p^J \tag{3.97}$$

$$\Sigma = A_J(R)\big[\cos(2\pi\arg 2)\cos^2(2\pi\Phi)\langle\Sigma|\Sigma_J\rangle\langle\Sigma_J|\Sigma\rangle \\ -\cos(2\pi\arg 2)\sin^2(2\pi\Phi)\langle\Delta|\Sigma_J\rangle\langle\Sigma_J|\Delta\rangle \\ -\sin(2\pi\arg 2)\cos(2\pi\Phi)\sin(2\pi\Phi) \\ \big(\langle\Delta|\Sigma_J\rangle\langle\Sigma_J|\Sigma\rangle + \langle\Sigma|\Sigma_J\rangle\langle\Sigma_J|\Delta\rangle\big)\big] \tag{3.98}$$

$$\Delta = A_J(R)\big[\cos(2\pi\arg 2)\cos^2(2\pi\Phi)\langle\Delta|\Sigma_J\rangle\langle\Sigma_J|\Sigma\rangle \\ -\cos(2\pi\arg 2)\sin^2(2\pi\Phi)\langle\Sigma|\Sigma_J\rangle\langle\Sigma_J|\Delta\rangle \\ -\sin(2\pi\arg 2)\cos(2\pi\Phi)\sin(2\pi\Phi) \\ \big(\langle\Sigma|\Sigma_J\rangle\langle\Sigma_J|\Sigma\rangle + \langle\Delta|\Sigma_J\rangle\langle\Sigma_J|\Delta\rangle\big)\big] \tag{3.99}$$

From the receiver I and Q processing the two complex samples are

$$\Sigma = A_J(R)e^{\left[2\pi i\left(\beta\Phi p - \frac{2R_J}{\lambda} + \eta_p^J\right)\right]}\big[\cos^2(2\pi\Phi)\langle\Sigma|\Sigma_J\rangle\langle\Sigma_J|\Sigma\rangle \\ -\sin^2(2\pi\Phi)\langle\Delta|\Sigma_J\rangle\langle\Sigma_J|\Delta\rangle \\ + i\cos(2\pi\Phi)\sin(2\pi\Phi) \\ \big(\langle\Delta|\Sigma_J\rangle\langle\Sigma_J|\Sigma\rangle + \langle\Sigma|\Sigma_J\rangle\langle\Sigma_J|\Delta\rangle\big)\big] \tag{3.100}$$

$$\Delta = A_J(R)e^{\left[2\pi i\left(\beta\Phi_p - \frac{2R_J}{\lambda} + \eta_p^J\right)\right]} \Big[\cos^2(2\pi\Phi)\langle\Delta|\Sigma_J\rangle\langle\Sigma_J|\Sigma\rangle$$
$$-\sin^2(2\pi\Phi)\langle\Sigma|\Sigma_J\rangle\langle\Sigma_J|\Delta\rangle$$
$$+i\cos(2\pi\Phi)\sin(2\pi\Phi) \quad (3.101)$$
$$\left(\langle\Sigma|\Sigma_J\rangle\langle\Sigma_J|\Sigma\rangle + \langle\Delta|\Sigma_J\rangle\langle\Sigma_J|\Delta\rangle\right)\Big]$$

The dominant terms are the terms quadratic in the Σ beam patterns and linear in difference patterns.

$$\Sigma = A_J(R)e^{\left[2\pi i\left(\beta\Phi_p - \frac{2R_J}{\lambda} + \eta_p^J\right)\right]} \left[\cos^2(2\pi\Phi)\langle\Sigma|\Sigma_J\rangle\langle\Sigma_J|\Sigma\rangle\right] \quad (3.102)$$

$$\Delta = A_J(R)e^{\left[2\pi i\left(\beta\Phi_p - \frac{2R_J}{\lambda} + \eta_p^J\right)\right]} \Big[\cos^2(2\pi\Phi)\langle\Delta|\Sigma_J\rangle\langle\Sigma_J|\Sigma\rangle$$
$$+i\cos(2\pi\Phi)\sin(2\pi\Phi) \quad (3.103)$$
$$\left(\langle\Sigma|\Sigma_J\rangle\langle\Sigma_J|\Sigma\rangle\right)\Big]$$

As in the previous section the EA terms that contribute to the monopulse ratio are given from (3.102) and (3.103) as

$$\frac{\Delta}{\Sigma} = i\tan(2\pi\Phi) + \frac{\langle\Delta|\Sigma_J\rangle}{\langle\Sigma|\Sigma_J\rangle} \quad (3.104)$$

From (3.67), this expression is approximately

$$\frac{\Delta}{\Sigma} = i\frac{d\pi}{\lambda} \cdot \psi \quad (3.105)$$

It is remembered that the angle in this case is the angle off ASM antenna bore sight to the EA system and not to the target.

3.6 Summary of Model

In sum the various expressions above can be combined with a receiver noise model term. The noise terms are assumed to be white noise. The classic processing for a single CPI results in two range Doppler arrays of data. At each range Doppler cell the data can, in general, have the following combinations of terms:

$$\Sigma = \Sigma_S + \Sigma_J + \Sigma_N \qquad (3.106)$$

$$\Delta = \Delta_S + \Delta_J + \Delta_N \qquad (3.107)$$

At times, a useful approximation for (3.107) comes from (3.89) and (3.105).

$$\Delta = iK\psi_S\Sigma_S + iK\psi_J\Sigma_J + \Delta_N \qquad (3.108)$$

As stated above for each CPI, an array of Σ channel data is collected together with an array of corresponding Δ channel data. In the search mode, each cell is examined via a CFAR type of algorithm to determine cells that may contain the target of interest. All of these potential cells are examined to estimate the cell (or cells) that most probably contain the target. The ASM switches to the track mode. The selected cell is made the focus of attention via the creation and execution of various standard track gates. The antenna is generally pointed in the direction of the target, and measurements are made for use in the guidance subsystem. The primary interest for the remainder of this text is the classification mode. The purpose of the classification mode is to determine the cell most likely containing target return. These modes can be understood as a form of hypothesis testing.

From Chapter 1, Strategy 1 consists of the HVU and a single decoy. Examining the full Σ channel range Doppler array of P samples, each cell contains just noise except the two cells that contain either the HVU or the decoy. It is assumed that these two cells shown in Figure 3.14 contain samples at a reasonably higher signal level than the remaining cells.

Thus, the detection of these two cells is assumed. (Keep in mind that the diagram represents a Doppler range array at a particular antenna pointing angle.) Each of the cells is tested as containing target data or containing decoy data. The classification task relates to the hypothesis test of each of these two range Doppler cells as one of the following options:

$$H0: \Sigma = \Sigma_S + \Sigma_N \qquad (3.109)$$

$$H1: \Sigma = \Sigma_J + \Sigma_N \qquad (3.110)$$

Strategy 2 consists of the HVU and an off-board platform using some form of EA to defend the HVU. The EA is some form of electronically generated

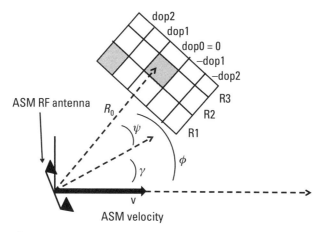

Figure 3.14 Strategy 1 Σ channel array showing two cells under test.

false target or targets. In addition, there may or may not be a decoy as the fleet desired target for the ASM. In this case, there may be multiple cells under test and the EA may change with time. The basic testing for each cell is the same as shown in (3.109) and (3.110) for cells containing a target of interest or an EA generated false target or a decoy (or a ship that is not the target of interest).

Strategy 3 consists of the HVU and/or another platform using cover jamming. In general the EA may include some combination of electronically generated false targets. In addition, there may or may not be a decoy as the desired target for the ASM. The testing in these latter cases is either (3.109) or (3.110).

In the case of cover noise EA, each cell contains significant amplitude and each cell contains EA noise jamming level. The goal is to determine the cell containing both the target return plus the jamming level. In this case the hypothesis testing for all cells is

$$H0: \Sigma = \Sigma_S + \Sigma_J + \Sigma_N \quad (3.111)$$

$$H1: \Sigma = \Sigma_J + \Sigma_N \quad (3.112)$$

Finally, it is noted for later use that the classical ASM sensor uses only the Σ channel data array for all functions except the use of a corresponding cell from the Δ channel to form the monopulse ratio for angle (line of sight relative to the antenna center line) estimation. In Chapter 6, it is shown that the modern ASM sensor uses the full ($2P$) array of data consisting of both the

Σ and Δ channel arrays. It seems intuitive that optimal two channel processing would result in better performance simply because twice as much data is processed.

Consider the data from the two receivers prior to the discrete Fourier transform (Doppler processing). Using (3.81) and (3.82), combine the Σ and Δ receiver returns to form two new combinations at each range sample in the range swath. Alternate the + and − components for each pulse p to form a single $2P$ component target echo vector for a CPI of duration PT. This new $2P$ vector at this range sample has components

$$X_{S+}(p) = A_S(R) \cdot e^{\left[2\pi i\left(\beta\Phi + f_D T\right)p - \frac{2R_0}{\lambda}\right]} e^{+2\pi i\Phi} S_+ \quad (3.113)$$

$$X_{S-}(p) = A_S(R) \cdot e^{\left[2\pi i\left(\beta\Phi + f_D T\right)p - \frac{2R_0}{\lambda}\right]} e^{-2\pi i\Phi} S_- \quad (3.114)$$

The following alternative expressions are included for completeness and use in the later chapters

$$S_+ = \cos(2\pi\Phi)\left(\langle\Sigma|\Omega|\Sigma\rangle + \langle\Delta|\Omega|\Sigma\rangle\right) \\ + i\sin(2\pi\Phi)\left(\langle\Delta|\Omega|\Delta\rangle + \langle\Sigma|\Omega|\Delta\rangle\right) \quad (3.115)$$

$$S_- = \cos(2\pi\Phi)\left(\langle\Sigma|\Omega|\Sigma\rangle - \langle\Delta|\Omega|\Sigma\rangle\right) \\ - i\sin(2\pi\Phi)\left(\langle\Delta|\Omega|\Delta\rangle - \langle\Sigma|\Omega|\Delta\rangle\right) \quad (3.116)$$

$$2S_+ = \exp(2\pi i\Phi)\left(\langle\Sigma|\Omega|\Sigma\rangle + \langle\Delta|\Omega|\Sigma\rangle + \langle\Sigma|\Omega|\Delta\rangle + \langle\Delta|\Omega|\Delta\rangle\right) \\ + \exp(-2\pi i\Phi)\left(\langle\Sigma|\Omega|\Sigma\rangle + \langle\Delta|\Omega|\Sigma\rangle - \langle\Sigma|\Omega|\Delta\rangle - \langle\Delta|\Omega|\Delta\rangle\right) \quad (3.117)$$

$$2S_- = \exp(2\pi i\Phi)\left(\langle\Sigma|\Omega|\Sigma\rangle - \langle\Delta|\Omega|\Sigma\rangle + \langle\Sigma|\Omega|\Delta\rangle - \langle\Delta|\Omega|\Delta\rangle\right) \\ + \exp(-2\pi i\Phi)\left(\langle\Sigma|\Omega|\Sigma\rangle - \langle\Delta|\Omega|\Sigma\rangle - \langle\Sigma|\Omega|\Delta\rangle + \langle\Delta|\Omega|\Delta\rangle\right) \quad (3.118)$$

Similarly, from (3.100) and (3.101), the alternative to the Σ and Δ outputs for the EA noise jamming at each range sample are the $2P$ component X vectors

$$X_{J+}(p) = A_J(R) \cdot e^{\left[2\pi i\left(\beta\Phi p - \frac{2R_J}{\lambda} + \eta_p^J\right)\right]} e^{+2\pi i\Phi} J_+ \quad (3.119)$$

$$X_{J_-}(p) = A_J(R) \cdot e^{\left[2\pi i\left(\beta\Phi p - \frac{2R_J}{\lambda} + \eta'_p\right)\right]} e^{-2\pi i \Phi} J_- \qquad (3.120)$$

$$J_+ = \cos(2\pi\Phi)\big[\langle\Sigma|\Sigma_J\rangle\langle\Sigma_J|\Sigma\rangle + \langle\Delta|\Sigma_J\rangle\langle\Sigma_J|\Sigma\rangle\big]$$
$$+ i\sin(2\pi\Phi)\big[\langle\Sigma|\Sigma_J\rangle\langle\Sigma_J|\Delta\rangle + \langle\Delta|\Sigma_J\rangle\langle\Sigma_J|\Delta\rangle\big] \qquad (3.121)$$

$$J_- = \cos(2\pi\Phi)\big[\langle\Sigma|\Sigma_J\rangle\langle\Sigma_J|\Sigma\rangle - \langle\Delta|\Sigma_J\rangle\langle\Sigma_J|\Sigma\rangle\big]$$
$$+ i\sin(2\pi\Phi)\big[\langle\Sigma|\Sigma_J\rangle\langle\Sigma_J|\Delta\rangle - \langle\Delta|\Sigma_J\rangle\langle\Sigma_J|\Delta\rangle\big] \qquad (3.122)$$

Thus, instead of the Σ and Δ channel arrays, one can process the full $2P$ component array for each range sample and CPI

$$X = X_S + X_J + X_N \qquad (3.123)$$

For Strategy 1, the equivalent to (3.109) and (3.110) are

$$H0: X = X_S + X_N \qquad (3.124)$$

$$H1: X = X_J + X_N \qquad (3.125)$$

For cover noise EA, the expressions equivalent to (3.111) and (3.112) are

$$H0: X = X_S + X_J + X_N \qquad (3.126)$$

$$H1: X = X_J + X_N \qquad (3.127)$$

3.7 Detection versus Classification and EP

As described above, the ASM typically has a priori information about the desired target including target type, RCS level, and location. During the ASM approach, it may climb above the radar horizon to update this information as it closes and flies selected waypoints. The final ascent above the radar horizon is considered the beginning of the ASM terminal phase. Once in its terminal phase the ASM has about 30s at most to detect the possible targets, select the proper target, and execute its guidance.

During the search mode or reacquisition mode, the detection of targets is generally successful unless cover noise EA is implemented from the onset of the ASM sensor radiating. Cover noise EA must eventually be generated by a source that is not in the line of sight of any targets or burn through may reveal the ship target location.

Once a target is selected the measurements during track mode are more than adequate to guide an autonomous ASM to this target. As a consequence, the goal of fleet defense EA must be to entice the ASM sensor to select and track an off-board decoy as the target. Hence the common phrase that the goal is to make the decoy look more like a target and to make the ship look less like a target. The modern EW information battle from the fleet EA perspective is a battle to make the off-board decoy seem to have the characteristics of the HVU and/or to hide or corrupt the characteristics of the HVU.

The critical function of the ASM sensor and its DSP is to protect the sensor from this possibility via sophisticated and rapid DSP EP algorithms. From the perspective of the ASM the goal of the EW information battle is to measure and test parameters from the various possible targets in the cells under test to optimally select the correct target by exploiting these target characteristics.

In sum, each range or range Doppler cell location in the array contains a pixel of magnitude (or magnitude and phase) data. The characteristics of this data must be used to conduct a hypothesis test to estimate whether or not the cell contains a target, the optimal target, deceptive data (false target or cover noise), or receiver noise. Having selected the cell of interest, its data contains measures of range, radial speed, and angle off bore sight. These measurements are available to the ASM for each CPI and sent continuously to the ASM guidance subsystem. The classification of the cell in the data array as the desired target is the critical function of target classification or EP. The several parameters available for this EP task include those listed in Table 3.5 and are

Table 3.5
Common Parameters for Target EP

Range and range statistics
Doppler and Doppler statistics
Range and Doppler image measures (sparseness and denseness)
Angle and angle statistics
Amplitude and amplitude statistics
Polarization information

discussed in subsequent chapters. The ultimate goal of the sensor processing is to guide the ASM to impact the desired ship target.

References

[1] Tsui, J., *Digital Techniques for Wideband Receivers*, Norwood, MA: Artech House, 2001.

[2] Chen, V. C., *The Micro-Doppler Effect in Radar*, Norwood, MA: Artech House, 2011.

[3] Schleher, D. C., *Electronic Warfare in the Information Age*, Norwood, MA: Artech House, 1999.

[4] Pace, P. E., *Detecting and Classifying Low Probability of Intercept Radar*, Norwood, MA: Artech House, 2009.

[5] Richards, M. A., *Fundamentals of Radar Signal Processing*, New York, NY: McGraw-Hill, 2005.

[6] Wehner, D., *High-Resolution Radar*, Norwood, MA: Artech House, 1995.

4
Extended Target EP Signal Processing

The remainder of the book describes modern, practical EP techniques made possible by a combination of state-of-the art radar technology with rapid digital signal processors. The classical radar sensor collecting ambiguous measurements is now replaced by a sensor that collects precise digital data and simultaneously applies multiple rapid and purposeful digital signal processing algorithms. With greatly improved detection and tracking capability, the primary objective of these algorithms is to ensure that the sensor is focused on the correct target, that is, to protect the sensor from being deceived into targeting a decoy.

In this chapter, several ASM radar EP techniques are described that specifically exploit the true nature of an extended naval target that is revealed through the statistical and physical properties of the HVU RCS measurement. Examples of these statistics include the mean, the variance, and the time correlation properties of the RCS. These practical EP techniques readily counter several present-day EA strategies that rely on the generation of any of a variety of false targets, including electronically generated false targets, reflector-based decoys, and chaff. The decoy, as well as any other false targets employed by the fleet, must successfully mimic the properties of the true ship target echo, whether the EA system is onboard one of the ships or on an off-board platform.

In each section, basic EA strategies employed by the fleet to counter the ASM terminal engagement phase are examined in more detail. The ultimate goal of each strategy is to deceive the ASM sensor into selecting and tracking an off-board decoy representing a false HVU target. The EA strategies may assume the ASM sensor chooses the off-board decoy at the onset of the engagement or not. If not, the EA strategy must include the means to suppress the choice of the true target during the engagement and/or to enhance the option of choosing the decoy.

In the first section, the properties are described of a false target electronically generated via a DRFM-based EA system. If this target is a seduction target, it must mimic true target features while also moving in a physically realistic manner in a multiple dimensional space. Another strategy made possible by the DRFM EA system exploits the generation of multiple false targets at various ranges and Doppler values to add confusion to the scene viewed by the ASM sensor to reduce the probability of the ASM sensor choosing the correct target. Various EP techniques are discussed that exploit the group characteristics of the targets. The DRFM affords the EA designer great flexibility in the generated characteristics of the false target. However, to be successful the EA designer must be aware of the capabilities available to the ASM sensor EP designer.

In the next section, the characteristics are described of several configurations of passive decoys used to generate a false target by reflective means. The reflector-based decoy is typically on a floating platform. Again, the properties of a passive decoy are compared to those of the true ship target. The advantages and disadvantages of the passive decoy are detailed.

A common form of passive false target in use since the 1940s is chaff. Several techniques have been published by People's Republic of China (PRC) personnel to specifically address the ASM EP against chaff. These techniques are described in Section 4.3. These PRC authored papers have introduced various metrics and terminology that have been generally adopted by the naval EW community. In this section, the underlying philosophy as well as various simple variations and implementations of these techniques is described. In addition, it is shown how the fleet may exploit the EP techniques as a part of the EA defense strategy.

In the final section, there is a discussion of the use of multiple coherent receivers to interrogate the properties of an extended target. The coherent use of dual receivers made possible monopulse angle measurement capability for improved guidance capability. In this section, it is shown that an extended target can be distinguished from a point false target by the statistics of its monopulse angle measurement. The monopulse measurement is the ratio of the coherent returns in the two ASM sensor receiver channels. The classic counter to the monopulse angle measurement requires the use of dual coherent source (DCS) jamming for angle deception. DCS techniques include cross-polarization jamming, cross-eye jamming, and terrain bounce plus the various combinations of these techniques. This section contains a brief summary of DCS EA as it may be used to better mimic the statistics of an extended target to counter this EP technique.

4.1 Target Classification: False Targets

The realistic premise of modern EW is that the modern radar guided ASM in the terminal phase of its engagement will detect the one or more naval targets within its chosen range swath. The sensor will use its prior information to aid its choice of the optimal target assumed to be the HVU. The sensor will then track this target, while perhaps monitoring multiple targets within its view. For improved performance the ASM will point the transmit antenna beam primarily in the direction of the chosen target throughout this phase of the engagement. Figure 4.1 illustrates the geometry (not drawn to scale).

If the ASM sensor is in track mode and pointing its antenna at the HVU, the HVU will be generally on the bore sight of the ASM antenna beam. This is termed hypothesis H0. If the ASM sensor is in track mode and pointing at the decoy (Hypothesis H1), the HVU will be off bore sight of the ASM antenna beam.

$$\text{H0: Angle to HVU} = \psi_S = 0 \quad (4.1)$$

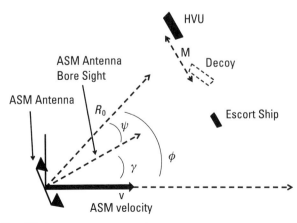

Figure 4.1 Simplified engagement geometry.

$$H1: \text{Angle to HVU} = \psi_s = \frac{M}{R_0} \qquad (4.2)$$

If the two targets (decoy and HVU) are close together in angle and nearly identical in feature, then the probability of the ASM choosing the decoy (H1) is expected to be about 0.5. One of several EA strategies may be employed by the fleet to improve its survivability. The ultimate goal of the fleet EA defense is to increase the probability that the ASM sensor chooses the decoy (H1) and/or to decrease the probability that the ASM sensor chooses the HVU (H0).

An example of a defensive strategy (Strategy 3) is to generate cover noise jamming to hide the HVU. The initial goal is to force the ASM sensor into the HOJ mode. This EA can be transmitted from the decoy, from the HVU, or from an escort ship. If the EA cover jamming is generated by the decoy, the goal is to keep the ASM sensor in HOJ until the ASM is close enough that it cannot successfully correct its attention onto a ship target. Burn through must not occur until the HVU angle in (4.2) is large enough to prevent the ASM from reacquiring the HVU and successfully maneuvering to the HVU. Modern ASMs are highly maneuverable. Figure 4.2 shows the ASM sensor locked onto the cover jamming from the decoy and the ASM flying in the direction of the decoy. In the lower left of the figure, the Doppler range array of data from a single CPI shows nothing but noise visible. With no other useful data, the monopulse measurement is used by the ASM to guide toward the decoy: the source of the EA.

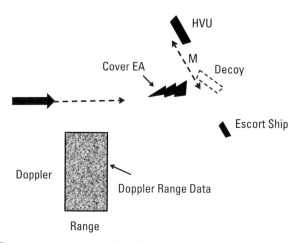

Figure 4.2 Decoy generated cover jamming.

Extended Target EP Signal Processing

In the other two cases (cover jamming from one of the ships), the strategy must include a quick transfer from the HOJ on the EA system to the decoy before burn through occurs, since the ASM sensor is pointed at a ship and the ASM is guiding toward a ship. If the decoy contains a DRFM-based EA system, an EA strategy may be for the decoy to generate a simulated false target to mimic burn through. Figure 4.3 illustrates this scenario just as the false target is generated.

Figure 4.4 illustrates the scenario after the ASM has switched to track mode on the false target generated by the decoy. With the antenna pointing

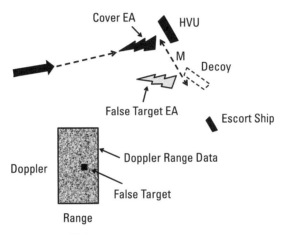

Figure 4.3 Decoy generated false target.

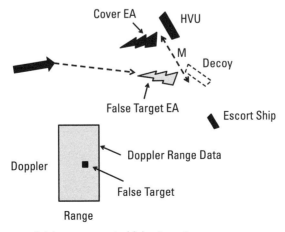

Figure 4.4 Successful decoy generated false target.

at the decoy, both the level of the HVU and the level of the cover jamming decrease.

A similar EA strategy was tested during an Advanced Technology Demonstration that culminated with a sea test on board a U.S. Navy ship in 1990. That strategy included a passive off-board decoy located close in angle to the HVU, but about 500m beyond the ship in range. The EA system onboard the HVU transmitted cover jamming that included a keeper pulse to simulate ship burn through. The keeper pulse was positioned about 500m beyond the ship in range. Figure 4.5 illustrates the geometry.

As mentioned previously, an ASM sensor in track mode generates narrow tracking gates to protect the sensor from interfering EA. Figure 4.6 shows the ASM sensor tracking the EA generated false target.

Once it is assumed that the ASM sensor is in track mode, the active EA from the HVU is discontinued. Since the tracking gates are narrow in range and Doppler while the antenna beam is about 10° wide, when the EA is discontinued there is still a target (the decoy) in the Doppler range gates of the ASM sensor. Thus, the ASM sensor continues to track as shown in Figure 4.7. Only now the ASM is tracking the decoy. As indicated, the cover EA as well as the EA-generated false target is discontinued.

As long as the ASM sensor is generally viewing in the direction of the HVU, there is the possibility of eventually detecting and choosing the HVU. It is imperative for the EA designer to set a goal of causing the ASM sensor to look away from the HVU as soon as possible in the engagement as indicated in these sample strategies. The most efficient EA method is to keep the ASM

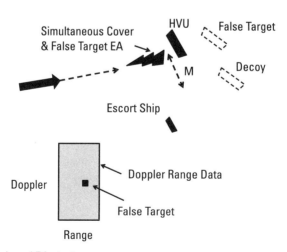

Figure 4.5 Onboard EA strategy.

Extended Target EP Signal Processing 121

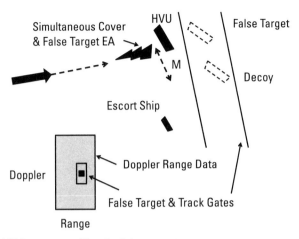

Figure 4.6 ASM sensor tracking the false target.

sensor in its track state, but to seduce the sensor into choosing a false target that separates in angle from the HVU.

As long as the ASM sensor is tracking an off-board decoy, it will not execute any tactics to find the HVU. Thus, there is an important requirement in defending the HVU to be able to generate false targets (electronic false targets and passive or active decoys) with characteristics similar to the real target (the HVU) at various ranges and Doppler frequencies. The focus of this text is on this critical task of target classification as ASM sensor EP. To be

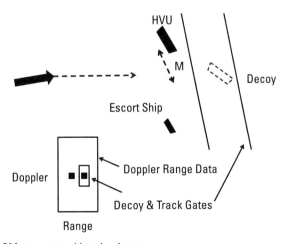

Figure 4.7 ASM sensor tracking the decoy.

successful, the EA engineer must be aware of the several characteristics of the HVU that can be readily measured by the ASM sensor in very few seconds.

The power density from the ASM radar transmission as observed at the target or the EA system is

$$\text{ASM Power density at EA} = \frac{P_A G_A(\psi)}{4\pi R^2} \qquad (4.3)$$

The level at the EA receiver is

$$\text{ASM Power at EA receiver} = \frac{P_A G_A(\psi) G_J \lambda^2}{(4\pi)^2 R^2} \qquad (4.4)$$

Using the expression for the power of a true target echo, the ASM will estimate the target RCS from (neglecting the coherent gain)

$$\text{Target Power received} = \frac{P_A [G_A(\psi)]^2 \sigma_T \lambda^2}{(4\pi)^3 R^4} \qquad (4.5)$$

Similarly, the power of the EA false target signal at the ASM radar is

$$\text{False Target Power received} = \frac{P_J G_A(\psi) G_J \lambda^2}{(4\pi)^2 R^2} \qquad (4.6)$$

Therefore, the ASM will estimate the RCS of a false target as

$$\sigma_T = \left(\frac{(4\pi)^2 R^2}{P_A G_A(\psi) G_J \lambda^2} \right) \cdot \frac{P_J G_J^2 \lambda^2}{4\pi} \qquad (4.7)$$

With the DRFM device, the EA system can store a high-fidelity copy of the sensor transmitted signal and then repeat it back with a variety of modifications such as amplitude as well as a variety of range and Doppler characteristics [1–3]. For the sample EA strategies discussed above to be successful, it is required the cover EA be high enough to cover the HVU ship and yet low enough that a realistic target might be visible. Figure 4.8 illustrates typical level requirements.

The EA systems are able to estimate the EA transmit power levels required to achieve an approximate RCS level at the ASM. It is expected that the ASM sensor knows the expected RCS of the HVU from prior intelligence

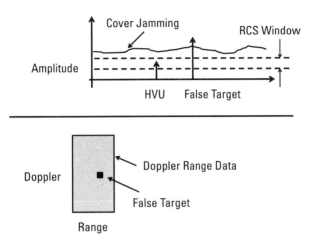

Figure 4.8 Example of amplitude of the ASM data.

information combined with prior measurements of the HVU if observations were made before the terminal phase. Thus, a typical first EP algorithm is to create a filter to only accept target measurements with RCS within a window of the expected or mean value.

Assume the HVU has RCS of about 40 dBsm. The cover jamming must be greater than this value to completely cover the HVU data. If the cover jamming level corresponds to an RCS level of 48 dBsm, then the ratio of the jamming level to the signal level (JSR) is 8 dB. The false target amplitude must correspond to a target level greater than the noise jamming level, for example, 55 dBsm, to be visible giving a signal level to noise jamming level of 7 dB for the false target sample. It is expected that the RCS acceptance window may be as small as 5 dB and only targets with RCS between 37.5 and 42.5 dBsm will be accepted. Thus, the false target in all of these strategies will be ignored by the ASM, and it will remain in HOJ until a viable option is observed. Additional EP counters to cover jamming strategies are discussed in detail in Chapter 6.

The first EP algorithm for the modern sensor is based on the mean of the target RCS. If the EA system or a decoy generates a false target with RCS that is too large or too small, it will be rejected. This is a simple and fundamental EP technique that must be countered by the EA system.

The most common false target generated by a DRFM-based EA system is commonly known in radar terminology as a Marcum target [4–6]. The Marcum target is represented as a single scattering element, and it is nonfluctuating in the sense that the complex echo can be represented as a constant amplitude echo plus Gaussian noise. As shown above, the mean level must be representative of a true target level corresponding to the RCS level for the ship class of the true target. From (4.6) and (4.7), it is seen that the correct level can be transmitted if the DRFM-based EA system successfully captures the signal transmitted from the ASM transmitter.

A simple and basic strategy (Strategy 1) is to consider the observed swath to contain the HVU and an off-board decoy. This decoy can be active in the sense that it contains a DRFM-based EA system that generates an electronic target. Or this decoy can be passive in the sense that the false target is generated by one or more reflectors. Passive decoys are discussed in the next two sections.

Assume that the active decoy generates an RCS equal to the expected RCS of the HVU. As the ASM sensor conducts its search mode and transitions to track mode, it is assumed that the probability of H0 or H1 is 0.5. If it is estimated (or assumed) that the ASM may be tracking the HVU, seduction EA may be employed (Strategy 2). For example, the HVU employs an onboard DRFM-based EA system to generate a false target in an attempt to seduce the track gates that may be on the HVU and to position these gates on the decoy. This requires the onboard EA system to know where the decoy is and to be able to estimate the range and Doppler location of the decoy within the ASM sensor Doppler range data array. It is assumed that the state may be as represented in Figure 4.9.

The EA system onboard the HVU attempts to generate a false target within the track (range and Doppler) gates located on the HVU for seduction or gate pull-off. This false target RCS must be larger than the HVU, but not so large that it exceeds the RCS acceptable window and is rejected. This false

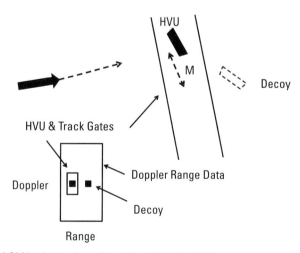

Figure 4.9 ASM in the track mode and tracking the HVU.

target must be moved from the Doppler and range of the HVU to the Doppler and range of the decoy. This must happen fast enough to prevent the ASM from impacting the HVU. And it must be slow enough so as not to significantly violate the physics of the properties of the false target as measured by the ASM sensor. (A target with range and Doppler that deviate from those of the true target is generally not physical.) Thus, the range and Doppler (and azimuth) are consistent with those of the HVU. This is illustrated in Figure 4.10.

The ASM knows its own closing speed and can estimate the target radial speed from its Doppler value. Given this information, the range track gate is updated from one CPI to the next by a filter approximated as

$$\hat{R}(t) = \hat{R}(t-T) - vT - v_T T \quad (4.8)$$

At this time, a measurement of the range is made

$$z(t) = R(t) + n(t) \quad (4.9)$$

Comparing the measurement with the update of the range gate leads to an innovations sequence that is random and not correlated with time. If, for example, the false target is moved to a slightly longer range, it would be at a range of

$$R_{FT}(t) = R(t) + \Delta R \quad (4.10)$$

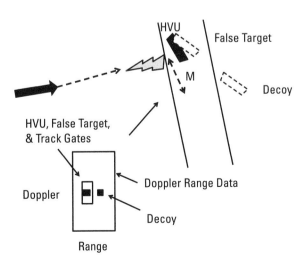

Figure 4.10 Seduction EA initiation.

To pull the false target away from the HVU and toward the decoy, the ΔR must be positive and constantly increasing. When the measurements of the range to the false target are compared to the filter update in (4.8), the resulting innovations sequence will be biased and not random. This can be detected by the ASM sensor.

Any of the standard EP techniques employed by radars to guard against track gate (range gate or velocity gate), pull-off may be implemented to prevent this state. A related and simpler test is a classic EP technique for coherent radar. As a result of the RCS window, the HVU and the false target must be close in amplitude. Assume that the HVU and false target are in a single Doppler range data array location at the onset of gate pull-off. From (3.81), the return in that location for pulse p is approximately

$$\Sigma(p) \approx A\left(e^{2\pi i f_D T p} + e^{2\pi i f_{FT} T p}\right) \qquad (4.11)$$

To be successful, the Doppler frequencies must be close. Define the sum and difference frequencies

$$f_+ = \frac{f_{FT} + f_D}{2} \qquad (4.12)$$

$$f_- = \frac{f_{FT} - f_D}{2} \qquad (4.13)$$

Then, the combined target and false target expression is

$$\Sigma(p) \approx 2A \cdot e^{2\pi i f_+ T p} \cdot \cos(2\pi f_- T p) \qquad (4.14)$$

So the combination of the ship target and false target attempting to seduce the track gate contains a slow amplitude modulation. The detection of this modulation is known as beat frequency detector (BFD) EP [4]. The upper portion of Figure 4.11 illustrates the ASM viewing the HVU and the false target geometry. At the lower left is an example of the Doppler filter response for a single CPI. As the two targets begin to separate in the unresolved Doppler filter, the mix of data with slightly differing frequency creates a beat frequency on the amplitude of the combined target illustrated in the center of the figure. By implementing a Fourier transform over several CPIs, the existence of the second (false target) is easily detected. This is illustrated in the lower right portion of the figure. False target velocity gate pull-off is readily detected by a coherent low probability of intercept (LPI) radar sensor.

Extended Target EP Signal Processing 127

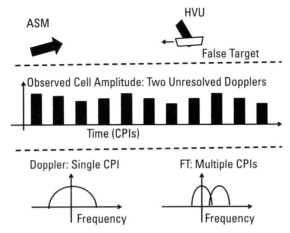

Figure 4.11 BFD example.

These classic gate pull-off EP techniques including the BFD technique are very robust. However, while they exploit the physics of the target motion, they do not exploit the RCS characteristics of an extended target. Therefore, they are not discussed any further in this work. There are many references that describe these algorithms, and it is expected that these algorithms are included in the DSP of the modern ASM sensor as required.

The ultimate goal of the EA strategy under consideration is to transition to the situation illustrated in Figure 4.12.

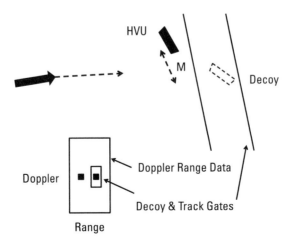

Figure 4.12 Goal of EA strategy.

The EA strategies described in this section seek to present the ASM sensor with an off-board decoy containing an active DRFM-based EA system that transmits a false target that successfully represents the features of the HVU. The false target described thus far is a nonfluctuating or Marcum target. The Marcum target is modeled as containing a single and constant scatter element or source.

Consider a simple object of length W that is composed of two identical scatter elements located at different positions within the unresolved range return of the ASM radar as shown in Figure 4.13.

The radar return of the object is

$$\Sigma \approx a \cdot \left\{ \cos\left[2\pi f\left(t - \frac{2(R_0 - \Delta R)}{c}\right)\right] + \cos\left[2\pi f\left(t - \frac{2(R_0 + \Delta R)}{c}\right)\right] \right\} \quad (4.15)$$

$$\Sigma \approx \left[2a \cdot \cos\left(2\pi f \frac{W \sin\theta}{c}\right)\right] \cdot \cos\left[2\pi f\left(t - \frac{2R_0}{c}\right)\right] \quad (4.16)$$

The amplitude of this return (the term in the first brackets on the right-hand side of the equation) varies with the angle θ. Figure 4.14 illustrates the variation of the RCS for a target consisting of two identical scatter elements. On the top-right, the object length is equal to the radar signal wavelength. On the bottom-left, the object length is equal to 3λ.

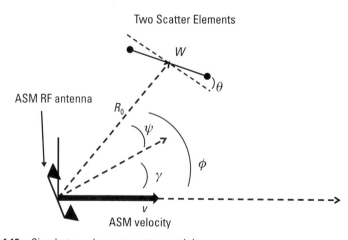

Figure 4.13 Simple two element scatter model.

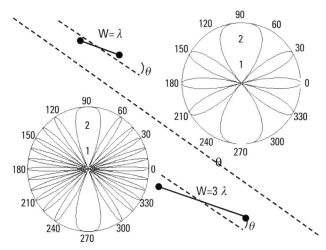

Figure 4.14 RCS of a simple two-scatter element rigid object.

In Figure 4.15, the rigid object consists of four identical scatter elements. The total length of the object is 3λ. The object contains two elements at the ends and two other elements on the object separated by λ.

Finally, the object consisting of the four identical scatter elements is assumed to be much longer than the radar wavelength (λ about 3 cm), and the four scatter elements are not colinear. The RCS as a function of viewing angle is seen in Figure 4.16 to quickly become quite complex.

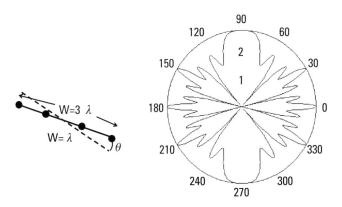

Figure 4.15 RCS of a simple four-scatter element rigid object.

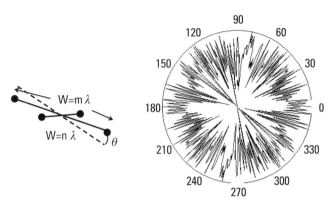

Figure 4.16 RCS for a complex four-scatter element rigid object.

As indicated, any slight change to the view of a complex target will cause the RCS to vary a significant amount. In addition, varying the carrier frequency enhances the fluctuations of the amplitude return from a complex, extended target (multiple unresolved scatter elements) as evident from (4.16). Thus, assuming that the range differences of the scatter elements are many multiples of the wavelength, this simple example demonstrates that the RCS is expected to be random from CPI to CPI, especially with frequency agility and time-varying geometry.

A simple estimate of the frequency requirements to decorrelate the RCS follows from (4.16). Decorrelation occurs when consecutive CPI measurements are 180° out of phase. Thus

$$\pi \le 2\pi f_1 \cdot \frac{W}{c} - 2\pi f_2 \cdot \frac{W}{c} \qquad (4.17)$$

$$\Delta f \ge \frac{c}{2W} \qquad (4.18)$$

For a naval target, W is as much as 50 to 70m even when viewed along the shorter dimension. This corresponds to the requirement for the frequency agility to be at least 2 to 3 MHz from one CPI to the next to decorrelate the consecutive RCS returns.

The HVU RCS is composed of many scatter elements spread over an extent much greater than the radar wavelength. As a result the RCS of a true target fluctuates over time. This fluctuation can be enhanced by frequency agility. The statistics of an extended (fluctuating) radar target are well known as Swerling statistics. The detection statistic functions are known for various types of complex objects. For each type of target, Swerling considered objects with very fast fluctuations and with slower fluctuations. Slow fluctuations can be considered to be roughly constant within a CPI but independent from one CPI to the next. Fast fluctuating targets can be considered to have RCS values independent from one pulse to the next. Table 4.1 gives the general characterization of the results.

For the purpose of this work, a complex naval target will have RCS that is generally independent from one CPI to the next. Figure 4.17 illustrates the concept of slow and fast fluctuations. The probability density functions for the different cases are Rayleigh for Cases I and II and Chi square for Cases III and IV.

The goal of this work is to consider fast practical EP algorithms for the radar sensor. HVU RCS is a combination of many unresolved scatter elements and the RCS or amplitude from a Swerling target is independent from one CPI to the next. A false target from a power-fixed EA system generates a

Table 4.1
Swerling Statistics Categories

Case	Fluctuations	Target
I	Slow	Many independent, similar elements
II	Fast	Many independent, similar elements
III	Slow	One dominant and many lesser elements
IV	Fast	One dominant and many lesser elements

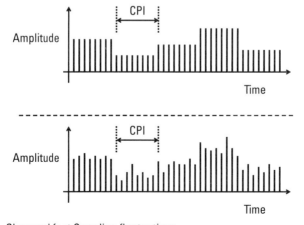

Figure 4.17 Slow and fast Swerling fluctuations.

much simpler Marcum target. Unless a more sophisticated DRFM scheme is employed, the false target statistics are much simpler than the RCS statistics of the HVU. From the previous chapter, the false target return for pulse p of a CPI is (with zero false target Doppler value)

$$\Sigma = A_J(R)e^{\left[2\pi i\left(\beta\Phi p - \frac{2R_J}{\lambda}\right)\right]}\Big[\cos^2(2\pi\Phi)\langle\Sigma|\Sigma_J\rangle\langle\Sigma_J|\Sigma\rangle \\ -\sin^2(2\pi\Phi)\langle\Delta|\Sigma_J\rangle\langle\Sigma_J|\Delta\rangle \\ +i\cos(2\pi\Phi)\sin(2\pi\Phi) \\ \left(\langle\Delta|\Sigma_J\rangle\langle\Sigma_J|\Sigma\rangle + \langle\Sigma|\Sigma_J\rangle\langle\Sigma_J|\Delta\rangle\right)\Big] \quad (4.19)$$

$$\Delta = A_J(R)e^{\left[2\pi i\left(\beta\Phi p - \frac{2R_J}{\lambda}\right)\right]}\Big[\cos^2(2\pi\Phi)\langle\Delta|\Sigma_J\rangle\langle\Sigma_J|\Sigma\rangle \\ -\sin^2(2\pi\Phi)\langle\Sigma|\Sigma_J\rangle\langle\Sigma_J|\Delta\rangle \\ +i\cos(2\pi\Phi)\sin(2\pi\Phi) \\ \left(\langle\Sigma|\Sigma_J\rangle\langle\Sigma_J|\Sigma\rangle + \langle\Delta|\Sigma_J\rangle\langle\Sigma_J|\Delta\rangle\right)\Big] \quad (4.20)$$

Again the leading terms are

$$\Sigma = A_J(R)e^{\left[2\pi i\left(\beta\Phi p - \frac{2R_J}{\lambda}\right)\right]}\Big[\cos^2(2\pi\Phi)\langle\Sigma|\Sigma_J\rangle\langle\Sigma_J|\Sigma\rangle\Big] \quad (4.21)$$

$$\Delta = A_J(R)e^{\left[2\pi i\left(\beta\Phi p - \frac{2R_J}{\lambda}\right)\right]}\Big[\cos^2(2\pi\Phi)\langle\Delta|\Sigma_J\rangle\langle\Sigma_J|\Sigma\rangle \\ +i\cos(2\pi\Phi)\sin(2\pi\Phi)\left(\langle\Sigma|\Sigma_J\rangle\langle\Sigma_J|\Sigma\rangle\right)\Big] \quad (4.22)$$

After Doppler processing, the return in the optimal Doppler range cell is

$$\Sigma = A_J(R)e^{-2\pi i\frac{2R_J}{\lambda}}\Big[\cos^2(2\pi\Phi)\langle\Sigma|\Sigma_J\rangle\langle\Sigma_J|\Sigma\rangle\Big] \quad (4.23)$$

$$\Delta = A_J(R)e^{-2\pi i\frac{2R_J}{\lambda}}\Big[\cos^2(2\pi\Phi)\langle\Delta|\Sigma_J\rangle\langle\Sigma_J|\Sigma\rangle \\ +i\cos(2\pi\Phi)\sin(2\pi\Phi)\left(\langle\Sigma|\Sigma_J\rangle\langle\Sigma_J|\Sigma\rangle\right)\Big] \quad (4.24)$$

Previously, it was shown that the ASM radar sensor processor only accepts returns corresponding to RCS values within an acceptable window. This exploits knowledge of the mean value of the target RCS. Simple and practical EP techniques have been developed that examine the higher order statistics of the RCS. Assume for simplicity that RCS for CPI n is

$$\sigma_n = \bar{\sigma} + \epsilon_n \qquad (4.25)$$

Consider the simple metric

$$\text{Metric} = \frac{\frac{1}{N}\sum_n (\sigma_{n+1} - \bar{\sigma}) \cdot (\sigma_n - \bar{\sigma})}{\frac{1}{N}\sum_n (\sigma_n - \bar{\sigma})^2} \qquad (4.26)$$

$$\langle \text{Metric} \rangle \approx C(1) \qquad (4.27)$$

The proposed metric is a measure of the Lag-1 term of the RCS autocorrelation function. It is expected that the Lag-1 term is close to 0 for HVU and close to 1 for false targets and simple decoys. A practical and simple estimate of the RCS Lag-1 correlation value can be readily computed by a set of simple running averages of the necessary terms estimated in the ASM DSP. These terms are combined at each CPI to estimate the Lag-1 correlation. Setting an appropriate threshold is adequate to estimate this property during target tracking in the terminal phase. For example, choose a filter gain (g) and for each estimate of σ measurement (and letting the minus indicate the prior [CPI] time sample and filter estimates)

$$S1 = S1_- + g \cdot (\sigma \cdot \sigma_- - S1_-) \qquad (4.28)$$

$$S0 = S0_- + g \cdot (\sigma \cdot \sigma - S0_-) \qquad (4.29)$$

$$M = M_- + g \cdot (\sigma - M_-) \qquad (4.30)$$

$$C(1) \approx \frac{S1 - M^2}{S0 - M^2} \qquad (4.31)$$

An example of experimental data is shown in Figure 4.18. As seen with a CPI of about 10 ms, the RCS of a ship target decorrelates in less than 1 CPI.

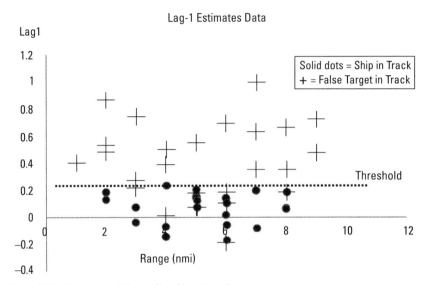

Figure 4.18 Experimental results of Lag-1 estimates.

False targets whose RCS do not decorrelate from one CPI to the next can be generally and quickly rejected by such an algorithm.

Personnel of the PRC have proposed exploiting a variety of target features to identify the HVU and the false target, including Doppler characteristics and antenna polarization in a variety of English journals [7–13]. As an example, consider the dominant terms for a true target from (3.84) and (3.85) and the corresponding dominant terms for a DRFM-generated false target repeated here as Equations (4.32) and (4.33) for the true target, and (4.34) and (4.35) for the false target

$$\Sigma = A_S(R) \cdot e^{\left[-2\pi i \frac{2R_0}{\lambda}\right]} \Big[\cos^2(2\pi\Phi)\langle\Sigma|\Omega|\Sigma\rangle - \sin^2(2\pi\Phi)\langle\Delta|\Omega|\Delta\rangle \\ + i\cos(2\pi\Phi)\sin(2\pi\Phi)\big(\langle\Delta|\Omega|\Sigma\rangle + \langle\Sigma|\Omega|\Delta\rangle\big) \Big] \quad (4.32)$$

$$\Delta = A_S(R) \cdot e^{\left[-2\pi i \frac{2R_0}{\lambda}\right]} \Big[\cos^2(2\pi\Phi)\langle\Delta|\Omega|\Sigma\rangle - \sin^2(2\pi\Phi)\langle\Sigma|\Omega|\Delta\rangle \\ + i\cos(2\pi\Phi)\sin(2\pi\Phi)\big(\langle\Sigma|\Omega|\Sigma\rangle + \langle\Delta|\Omega|\Delta\rangle\big) \Big] \quad (4.33)$$

$$\Sigma = A_J(R) e^{-2\pi i \frac{2R_J}{\lambda}} \Big[\cos^2(2\pi\Phi)\langle\Sigma|\Sigma_J\rangle\langle\Sigma_J|\Sigma\rangle \Big] \quad (4.34)$$

$$\Delta = A_J(R)e^{-2\pi i \frac{2R_J}{\lambda}} \left[\cos^2(2\pi\Phi)\langle\Delta|\Sigma_J\rangle\langle\Sigma_J|\Sigma\rangle \right. \\ \left. + i\cos(2\pi\Phi)\sin(2\pi\Phi)\big(\langle\Sigma|\Sigma_J\rangle\langle\Sigma_J|\Sigma\rangle\big) \right] \quad (4.35)$$

The several references discuss algorithms based on ASM radar with amplitude monopulse, whereas this work uses a model with phase monopulse. However, the gist of the arguments can be understood with the following observations. Consider first the target samples. Form the two-dimensional vector components for the sum Y^+ and difference Y^- components and gather common terms

$$Y_S^+ = \left[A_S(R) \cdot e^{-2\pi i \frac{2R_0}{\lambda}} \cdot \cos(2\pi\Phi) \right] \\ \cdot \left\{ [2\cos(2\pi\Phi) + i\sin(2\pi\Phi)]\langle U + i\sin(2\pi\Phi)\langle L \right\} \cdot (\Omega\Sigma\rangle) \quad (4.36)$$

$$Y_S^- = \left[A_S(R) \cdot e^{-2\pi i \frac{2R_0}{\lambda}} \cdot \cos(2\pi\Phi) \right] \\ \cdot \left\{ [2\cos(2\pi\Phi) - i\sin(2\pi\Phi)]\langle U - i\sin(2\pi\Phi)\langle L \right\} \cdot (\Omega\Sigma\rangle) \quad (4.37)$$

The first term in brackets on the right-hand side is complex amplitude and a cosine angle common to both the components. The second terms are two-dimensional polarization beam pattern vectors that depend only on the ASM antenna properties at this look angle. The third terms represent the source properties, including polarization. The authors argue that given the two Y components and knowing the second terms, the equations can be used to solve the source polarization information

$$\text{Signal Source} \equiv \Omega\Sigma\rangle \quad (4.38)$$

To understand this conjecture form the Jones vector approximation to the receive beam vectors

$$\langle U \approx \begin{bmatrix} 1 & \rho_U \end{bmatrix} \quad (4.39)$$

$$\langle L \approx \begin{bmatrix} 1 & \rho_L \end{bmatrix} \quad (4.40)$$

Next form the Y vector from (4.36) and (4.37)

$$Y_S = k \cdot B \cdot \Omega \Sigma \rangle \tag{4.41}$$

$$B = \begin{bmatrix} 2[\cos(2\pi\Phi) + i\sin(2\pi\Phi)] & [2\cos(2\pi\Phi) + i\sin(2\pi\Phi)]\rho_U + i\sin(2\pi\Phi)\rho_L \\ 2[\cos(2\pi\Phi) - i\sin(2\pi\Phi)] & [2\cos(2\pi\Phi) - i\sin(2\pi\Phi)]\rho_U - i\sin(2\pi\Phi)\rho_L \end{bmatrix} \tag{4.42}$$

Consider now the false target expressions. There is a common term that represents the signal received by the EA system. This term can be included as in the EA transmit amplitude and included in the false target amplitude. Therefore

$$A' \equiv A_J(R) e^{-2\pi i \frac{2R_J}{\lambda}} \cdot \cos(2\pi\Phi) \cdot \langle \Sigma_J | \Sigma \rangle \tag{4.43}$$

For the remaining bracket, antenna polarization terms include only the parallel polarization terms. The two Y components for the false target are thus

$$Y_J^+ = A' \cdot \{[2\cos(2\pi\Phi) + i\sin(2\pi\Phi)]\langle U + i\sin(2\pi\Phi)\langle L\} \cdot (\Sigma_J\rangle) \tag{4.44}$$

$$Y_J^- = A' \cdot \{[2\cos(2\pi\Phi) - i\sin(2\pi\Phi)]\langle U - i\sin(2\pi\Phi)\langle L\} \cdot (\Sigma_J\rangle) \tag{4.45}$$

In this case the source is identified as

$$\text{EA Source} \equiv \Sigma_J \rangle \tag{4.46}$$

The Y vector in this case is

$$Y_J = k' \cdot B \cdot \Sigma_J \rangle \tag{4.47}$$

Thus, to solve the polarization of the target source in (4.41) for the HVU or (4.47) for the false target requires that B have an inverse. Then the solution is

$$\text{Polarization vector} = \text{constant} \cdot B^{-1} \cdot Y \tag{4.48}$$

The inverse exists if the determinant of B is nonzero. From the definition of B, the determinant is

$$|B| = 4i\cos(2\pi\Phi) \cdot \sin(2\pi\Phi) \cdot (\rho_U - \rho_L) \tag{4.49}$$

Therefore, the inverse exists and the task is solvable if the two subantennas have some nonzero cross-polarization components and the two subantennas are not perfectly matched in these components. As stated in the reference, the antennas have some cross-polarization components at most angles and real antennas are not perfectly matched. It is assumed by the authors that these components can be measured or estimated and the polarization of the source can be measured, revealing its identity. For the general model being considered, the ASM antenna is assumed to be linearly polarized, whereas the EA antenna is most commonly circularly polarized. The argument is, therefore, made that the ASM can identify echoes from real targets versus false targets from an EA system by observing the source polarization.

The scenario of this section is the ASM sensor viewing a range swath consisting of an HVU ship target and an off-board false target generated by a DRFM-based EA system. It is now further assumed that the off-board EA system generates an array of multiple false targets at a variety of Doppler values and ranges. The goal of this strategy (Strategy 2) is to confuse the ASM sensor and increase the probability that the ASM sensor selects one of the false targets. This correspondingly decreases the probability of the ASM sensor selecting the HVU.

For this class of EA, it is conjectured by several PRC engineers that many characteristics of the false targets are correlated within a CPI and over time [10]. These characteristics include RCS statistics as well as monopulse ratio statistics. It is especially realized that all of the false targets emanate from an identical antenna. Therefore, the source polarization measurements will be correlated.

Assuming the parameter $x^k(t_n)$ for targets k and $k + 1$ at the CPI t_n, the target pairwise correlation function estimates are formed as

$$C_k \approx \frac{\sum_n x^k(t_n) \cdot x^{k+1}(t_n)}{\sqrt{\left(\sum_n x^k(t_n) \cdot x^k(t_n)\right) \cdot \left(\sum_n x^{k+1}(t_n) \cdot x^{k+1}(t_n)\right)}} \quad (4.50)$$

This function is large for all target pairs except the two pairs containing the true target. This expression has been tested for the RCS parameter as well as the monopulse angle error. This EP approach has been observed to be very robust in quickly identifying the true target via approximations similar to (4.26) through (4.31).

In addition, after Doppler processing the monopulse ratio is

$$\frac{\Delta}{\Sigma} = i\tan(2\pi\Phi) + \frac{\langle\Delta|\Sigma_J\rangle}{\langle\Sigma|\Sigma_J\rangle} \qquad (4.51)$$

It is noted that the second term on the right-hand side of this equation is nonzero because of imperfections in the matching of the left and right subantennas. Personnel of the PRC have proposed exploiting this known feature when presented with multiple false targets [7–10]. An array of multiple false targets must have pairwise noncorrelating parameters, for example, RCS, monopulse, or these ASM sensor EP techniques will quickly identify the real target.

It is noted that the high-fidelity false targets are typically at a range greater than the range of the EA system. When generating a shorter range false target, the EA system will probably miss one or more of the LPI pulses; especially the first pulse in the CPI assuming frequency agility is implemented. Missing one or more pulses in the CPI results in a lower than planned RCS as well as the generation of ghost images peaking at the same range as the false target, but with a different Doppler value. These ghost images can be detected alerting the ASM processor to suspected false targets. For example, Figure 4.19 illustrates the case of missing every other pulse. The EA generated false targets must contain all of the pulses of the CPI. This difficult task is virtually impossible for false targets at a range shorter than the range of the EA system.

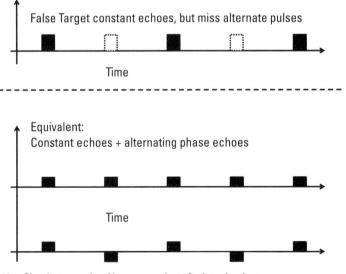

Figure 4.19 Simple example of interpretation of missed pulses.

Extended Target EP Signal Processing

In this simple example, it is assumed that the time (Doppler) portion of the echo from the target is constant (zero Doppler). From the expression in Chapter 3

$$\Sigma = A_S(R) \cdot e^{\left[2\pi i \left\{(\beta\Phi + f_D T)p - \frac{2R_0}{\lambda}\right\}\right]} \left[\cos^2(2\pi\Phi)\left(\left|g_\Sigma^p(\psi)\right|\right)^2 \sigma_{pp}\right] \quad (4.52)$$

As shown in the figure, a constant level with missing alternating pulses is equivalent to two targets of half level each. The first target is at the constant level (zero Doppler). The second target has a Doppler value corresponding to alternating sign (nonzero Doppler). An example is shown in Figure 4.20.

It was noted again that the Doppler measurement is a combination of target radial speed and angle off ASM antenna bore sight. Thus, some limited EP probing can be performed by slightly oscillating the antenna and viewing the aggregate motion of peaks in contiguous range Doppler array cells. Moving the antenna alters the angle off bore sight of the separate scatter elements causing them to move in varying patterns. If both the scatter elements are from the EA system (false target elements), they move in unison since the angle to each remains the same. If the scatter elements are at slightly differing angles (as from a true HVU), they move at differing phase. This technique will become more feasible as the Doppler resolution improves in the future. This technique is examined in detail in the following chapter. Figure 4.21 illustrates the effect.

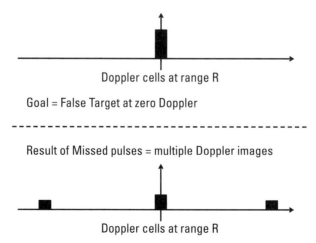

Figure 4.20 Image artifacts resulting from missed pulses.

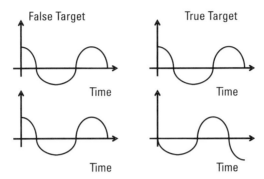

Figure 4.21 Antenna probe technique.

4.2 Target Classification: Decoys

Another type of false target is designated as a passive false target or passive decoy. A passive false target is generated via reflection. For example, the U.K.-manufactured Rubber Duck consists of one or more floating platforms tethered together. Each platform contains an array of one or more corner reflectors. The goal is to generate a target with time-varying RCS in the manner of the discussion of the previous section.

Attempts have been made to emulate HVU characteristics by adding multiple distributed reflecting elements and by cabling multiple decoys in tandem to achieve the proper length and width. The extent of the HVU in the Doppler range array is discussed in detail in the following section. For now, it is observed that the same characteristics as above can be utilized to identify passive decoys, especially lag-1 correlation techniques.

The next figures illustrate the RCS measurements together with the Lag-1 estimates, and the full autocorrelation function for typical ship targets compared with a typical decoy. Passive false targets are much more correlated from CPI to CPI than true ship targets. Figure 4.22 is the previous Lag-1 empirical data for ship targets and active EA generated false targets with the empirical estimates from a passive decoy added as the triangles. It is seen that the Lag-1 values of the RCS of the passive decoy are again significantly different from those of a ship target.

Figures 4.23 and 4.24 present empirical estimates of the full autocorrelation functions for both passive targets and true targets. The differences are obvious. In both the figures, the ship target RCS estimate is very close to

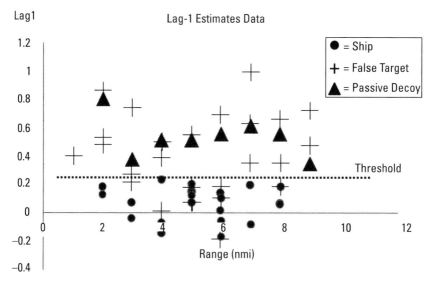

Figure 4.22 Experimental results of Lag-1 estimates.

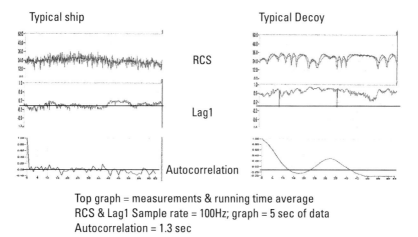

Top graph = measurements & running time average
RCS & Lag1 Sample rate = 100Hz; graph = 5 sec of data
Autocorrelation = 1.3 sec

Figure 4.23 Experimental results of correlation function estimates for a ship and a decoy.

24 dBsm and is quite noisy. The passive decoy RCS estimate in Figure 4.23 is close to 24, but quite smooth (time correlated). In Figure 4.24, the RCS is again smooth and a little high. The Lag-1 estimates for the ship targets are always 0.2 or lower. The Lag-1 estimates for the passive decoys are about 0.6 or larger. As an illustration the figures contain a snapshot (1.3s) of the full

142   Electronic Warfare Signal Processing

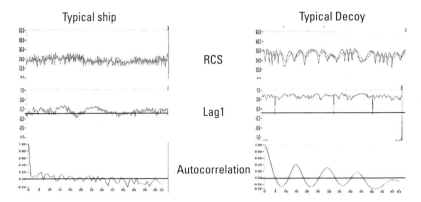

Top graph = measurements & running time average
RCS & Lag2 Sample rate = 100Hz; graph = 5 sec of data
Autocorrelation = 1.3 sec

Figure 4.24 Additional experimental results of correlation function estimates for ship and decoy.

autocorrelation function estimate. The autocorrelation functions for the real targets have the typical shape of a wideband random parameter. The same functions for the decoys are typical of a narrow-band (correlated) parameter.

4.3 Target Classification: Chaff

Another classic passive false target decoy is chaff. Chaff has been used as an EA to counter radar since the 1940s. Shortly after chaff was first introduced, operators observing crude radar displays began to ascertain features to distinguish chaff from true targets.

Chaff can be easily distinguished by an ASM pulsed Doppler radar by several of its characteristics. The first test for the true target is the RCS estimate as discussed above. The RCS from naval chaff may exceed the RCS of the HVU.

The second distinguishing features are the Doppler range image characteristics. Smaller ships occupy a single pixel (range/Doppler cell) of the modern ASM sensor, especially when viewed broad side. Many typical ASM pulsed Doppler radars can resolve the HVU echo responses into multiple range pixels. The HVU occupies several range cells even when viewed broad side. The HVU image is said to have hotspots in the sense that there are noncontiguous range cells that contain the highest RCS values. In addition, since the HVU is basically a rigid body and the resolution of the ASM radar in Doppler is rather coarse, all of the range echoes from the HVU occupy the same Doppler cell values.

Shortly after launch the chaff cloud spreads to multiple range pixels. As the chaff spreads, it is seen to have significant echo energy in the multiple contiguous range cells. The manner of the spreading (because of initial conditions and wind effects) causes the echoes in the contiguous range cells to be at various Doppler values. Thus, the chaff cloud typically has a fairly uniform response in multiple adjacent range cells, but at varying and somewhat random Doppler values especially at higher sea states. The image characteristics for real targets and chaff are illustrated in Figure 4.25.

Thus, a simple EP measure of chaff is to use range length and Doppler length, where length is the extent of the various range or Doppler values exceeding a threshold RCS value.

Figure 4.26 is an example of naval chaff being launched in close proximity to the ship. In the top sequence, the chaff is seen to spread in range and Doppler as it separates from the ship. The ship echo occupies a single Doppler range cell. The two images on the bottom row show the chaff at a later time. The chaff cloud has persisted for several minutes as it spreads and separates from the ship. The ship (at a different portion of the Doppler range array) still occupies a single cell.

This simple EP metric has been found to be very successful at identifying the chaff target. Since chaff has been used for so long, many works have been published to quantify these observations with a variety of more or less complicated metrics. Several works authored by PRC personnel have introduced various metrics and terminology that is being adopted by EW practitioners. For

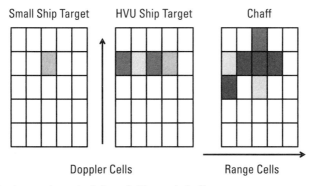

Figure 4.25 Image characteristics of ships and chaff.

example, the array dimensions of Doppler and range are considered separately. A random distribution of noncontiguous hot spots is termed sparse. Similarly, a clumping of contiguous hotspots is termed dense [14]. These definitions are combined with Table 4.2 to give Table 4.3

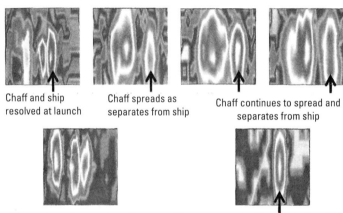

Chaff: 7 Doppler cells [vert] ´ 21 range cells [hor]
· Ship and chaff easily resolved
· EP = Use of range length and Doppler length (Sparseness/Denseness)

Chaff and ship resolved at launch

Chaff spreads as separates from ship

Chaff continues to spread and separates from ship

Chaff imaged on subsequent run: Range and Doppler characteristics of chaff and ship

·Note: Ship indicated by arrow

Figure 4.26 Chaff range/Doppler data.

Table 4.2
Image Lengths

	Ship	HVU	Chaff
Range length	0	>0 (distinct)	0 (contiguous)
Doppler length	0	0 (same)	>0 (distinct)

Table 4.3
Image Characteristics

	HVU	Chaff
Range	Sparse	Dense
Doppler	Dense	Sparse

Another way to view these definitions is to relate dense as indicating a nonrandom selection of values and sparse as a truly random selection of values. In the literature, these values are defined in the following manner. Observe the echo amplitude values in one dimension (range or Doppler). Convert the set of N data values to either 0 or 1, depending on whether it is below or above the median. Form the data subsets D_n as in (4.53)

$$D_n = \{x_k | \text{for } k = 1,\ldots,n;\ x_k = 0 \text{ if } < \text{median, else } x_k = 1\} \quad (4.53)$$

Let $m_N = N/2$ and let m_n = number of 1's in the subset D_n. The sparse discrepancy is defined as

$$d_n = \left| \frac{m_n}{m_N} - \frac{n}{N} \right| \quad (4.54)$$

The mean of this parameter is compared with a threshold. It can be seen that these are more or less measures of randomness when setting the median and investigating the order of values in range and/or Doppler directions.

A second definition of sparseness is more intuitive. In this definition, the sparseness r is the number of times the data in set D_N changes value. The statistics of r is known

$$\text{Mean}(r) = \frac{N+2}{2} \quad (4.55)$$

$$\text{Var}(r) = \frac{N}{4} \cdot \frac{N-2}{N-1} \quad (4.56)$$

A threshold and performance for this test can be modeled assuming a Gaussian distribution. Examining these statistics for a small set illustrates the behavior. Consider a random or sparse set (Set 1) and a nonrandom or dense set (Set 2)

Set 1 $\quad D_{10} = \{1,1,0,1,0,1,0,0,1,0\} \quad$ Sparse $\quad (4.57)$

Set 2 $\quad D_{10} = \{0,0,1,1,1,1,1,0,0,0\} \quad$ Dense $\quad (4.58)$

The results for these alternative definitions when applied to the hypothetical data sets are shown in Table 4.4.

Table 4.4
Alternative Definitions for Dense and Sparse Sets

	Set 1	Set 2
Definition 1	0.11	0.13
Definition 2: $r =$	7	2
	[Mean = 6; Var = 2.22]	

An important note must be made when it is realized that chaff can be easily distinguished from true targets. As emphasized before, the EW battle occurs very quickly. This is especially true in the airborne EW battle and the terminal phase of the naval battle with waves of ASM attacking the fleet. Since it is relatively easy for the ASM processor to detect and reject the chaff, a potentially useful EA strategy is to use a DRFM-based EA system to electronically generate what appears to be chaff at the true target (HVU) location in Doppler and range. This will prompt the ASM to reject this cell region since it has chaff characteristics instead of HVU characteristics. This can afford the fleet defense system valuable time during the terminal portion of the engagement and enhance its capability to generate more likely false targets or decoys.

4.4 Dual Coherent Source EA

In the previous chapter, a model of a simple scatter element was proposed. For a simple target, the measurements in the two receivers are

$$\Sigma = A_S(R) \cdot e^{\left[-2\pi i \frac{2R_0}{\lambda}\right]} \left[\cos^2(2\pi\Phi) \left(\left| g_\Sigma^p(\psi) \right| \right)^2 \sigma_{pp} \right] \quad (4.59)$$

$$\Delta = A_S(R) \cdot e^{\left[-2\pi i \frac{2R_0}{\lambda}\right]} \left[\cos^2(2\pi\Phi) g_\Delta^{p*}(\psi) \cdot g_\Sigma^p(\psi) \cdot \sigma_{pp} \right. \\ \left. + i\cos(2\pi\Phi)\sin(2\pi\Phi) \left(\left| g_\Sigma^p(\psi) \right| \right)^2 \sigma_{pp} \right] \quad (4.60)$$

It was noted that it is usual to form the monopulse ratio

$$\frac{\Delta}{\Sigma} = i\tan(2\pi\Phi) + \frac{g_\Delta^{p*}(\psi)}{g_\Sigma^{p*}(\psi)} \approx i\left(\frac{\pi d}{\lambda}\right)\psi \quad (4.61)$$

This ratio is the standard means used to estimate the angle off antenna bore sight of the target scatter element for a phase monopulse system. It is ideally a purely imaginary angle value plus an antenna term plus a noise term.

In the more general case, assume the target is composed of a multitude of scatter elements extending L length in range and W width in the azimuth direction, as illustrated in Figure 4.27. Assume these terms are within the same range Doppler cell. Each scatter element term in the two received signals is designated with the subscript k. The element of scatter amplitude a_k is at location W_k in azimuth relative to the nominal azimuth angle and $R_0 + \delta R_k$ in range relative to the nominal range.

The total return dominant terms can be approximated from the expressions in the previous chapter with the following modifications. Let

$$\delta\Phi_k = \text{ang}_k \cdot \frac{W}{R_0} \quad (4.62)$$

$$\text{ang}_k = \frac{d}{2\lambda} \cdot \frac{W_k}{W} \quad (4.63)$$

$$\text{term1} = \sum a_k \cdot e^{2\pi i \frac{2\delta R_k}{\lambda}} \cdot \langle \Sigma | \Omega_k | \Sigma \rangle \quad (4.64)$$

$$\text{term2} = \frac{2\pi W}{R_0} \cdot \sum a_k \cdot e^{2\pi i \frac{2\delta R_k}{\lambda}} \cdot \langle \Sigma | \Omega_k | \Sigma \rangle \cdot \text{ang}_k \quad (4.65)$$

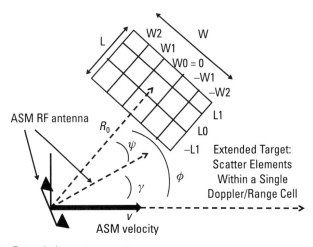

Figure 4.27 Extended target geometry.

Thus, for an extended object the dominant expressions within a particular range Doppler cell are

$$\Sigma = e^{\left[-2\pi i \frac{2R_0}{\lambda}\right]} \cos^2(2\pi\Phi) \cdot \text{term1} \tag{4.66}$$

$$\Delta = e^{\left[-2\pi i \frac{2R_0}{\lambda}\right]} i\cos(2\pi\Phi)\sin(2\pi\Phi) \cdot \text{term1} \\ + e^{\left[-2\pi i \frac{2R_0}{\lambda}\right]} \cos^2(2\pi\Phi) \cdot \text{term2} \tag{4.67}$$

Adding receiver noise the monopulse ratio (after normalization) is approximately an angle term, a target width term, and a noise term

$$z = \text{norm} \cdot \frac{\Delta}{\Sigma} = i\psi + (\alpha + i\beta) + (a + ib) \tag{4.68}$$

where

$$a + ib = \text{noise term} \tag{4.69}$$

$$\alpha + i\beta = \frac{i \cdot \text{term2}}{\text{term1}} \approx \frac{W}{R_0} \tag{4.70}$$

The standard monopulse noise terms are random variables with mean zero and standard deviation (square root of the variance) proportional to the antenna beam width and proportional to the square root of the inverse signal-to-noise ratio. From [5, 6, 10]

$$\sqrt{\langle a^2 + b^2 \rangle} = \frac{\theta_{BW}}{1.885 \cdot \sqrt{\text{SNR}}} \tag{4.71}$$

For a pulse Doppler radar the radar range equation is

$$\text{SNR} = \left[\frac{P_{pk} G^2 \lambda^2 N_{coh}}{(4\pi)^3 (kT_0) \text{BW} \cdot F \cdot L}\right] \cdot \frac{\sigma}{R_0^4} \tag{4.72}$$

From (4.71) and (4.72), it is seen that the receiver noise variance terms vary as the range to the target to the fourth power. This term decreases rapidly during the engagement.

The additional noise-like terms $(\alpha + i\beta)$ result from the extended and complex nature of the HVU and add an increasing (with decreasing range) complex part to the monopulse measurements. The real part of this term is called the Null Depth term and is commonly used to identify an extended target.

These additional noise-like terms are random variables with mean zero and variance proportional to the square of the ratio of target width over the range. This contribution to the monopulse measurement variance increases as the inverse range squared during the engagement. The exact value of the proportionality depends on the complex structure of the extended target (HVU). As described in [1, 5], a good approximation is

$$\sqrt{\langle \alpha^2 + \beta^2 \rangle} = \sqrt{2} \cdot 0.2 \cdot \frac{W}{R_0} \qquad (4.73)$$

Thus, the monopulse measurements from an extended (HVU) target have a larger variance than the monopulse measurements from point targets. And the variance difference between a real HVU and a false target increases with decreasing range.

Typical values for the monopulse variance can be estimated for a standard ASM pulse Doppler radar sensor used throughout this work. Figure 4.28 illustrates the expected performance. At long range the monopulse variance for both the targets cannot be readily distinguished. By a closing range of about 15 km, the difference between the variances is greater than 6 dB in magnitude and is readily distinguished.

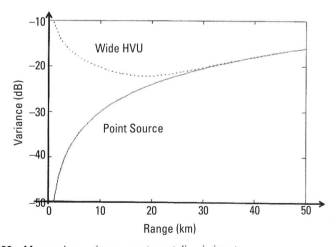

Figure 4.28 Monopulse variance as a target discriminant.

Figures 4.29 and 4.30 illustrate some test data showing the ability to discriminate ships from decoys using the monopulse variance. Figure 4.29 contains several test runs of ship data and several runs with DRFM-based EA false targets. Dashed lines have been added to qualitatively indicate the trends.

The same data has been presented in Figure 4.30. In addition, data from a passive decoy has been included. It is seen that the decoy variance appears to be more realistic than the DRFM-based EA false targets. This is expected, since the passive decoy consists of more than one scatter element while the EA false target is generated by a point source.

To counter this EP technique from an EA jamming system requires the capability to generate deceptive angular measurements against a monopulse radar. The usual way to generate angular deception against a monopulse radar and, thus, simulate these statistics from a jamming source is with DCS jamming techniques [4, 5, 15]. These jamming systems can be retrodirective or other variations, but must include two coherently controlled sources.

The outputs of the two EA transceivers for a retrodirective EA system are proportional (amplitude and phase) to the signals received in the opposite channel. Assume that the output of antenna 2 is proportional to amplitude

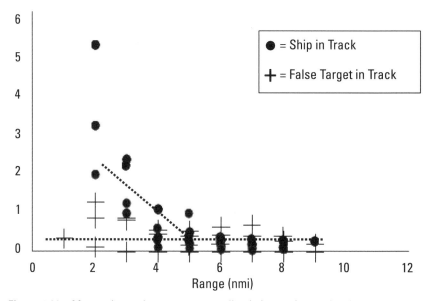

Figure 4.29 Monopulse variance as a target discriminant of an active decoy.

Monopulse Variance Estimates Data

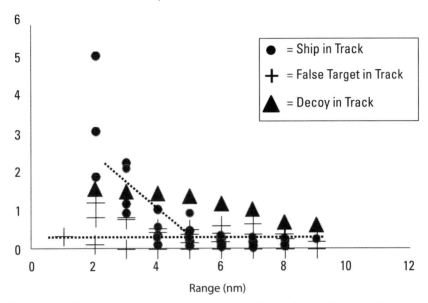

Figure 4.30 Monopulse variance as a target discriminant of an active or passive decoy.

(a) times the time history received at antenna 1. The output of antenna 1 is proportional to complex amplitude (az) times the input of antenna 2.

From the previous chapter, assume the signal is received at the two EA antennas and resent as shown in Figure 4.31. From the definition of DCS EA and the control parameter (z) the total EA transmit signal is

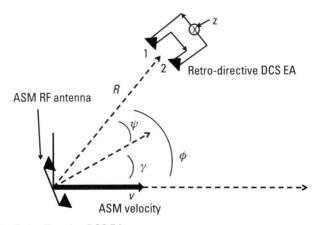

Figure 4.31 Retrodirective DCS EA.

$$\arg 1 = f_0 t + \delta f t + \frac{vt\cos(\varphi)}{\lambda} + \frac{v_T t \cos(\theta_T)}{\lambda} - \frac{R_0}{\lambda} \quad (4.74)$$

$$\begin{aligned}&1\rangle \cdot a \cdot z \cdot \big[\langle 2|\Sigma\rangle \cos(2\pi\arg 1)\cos(2\pi\Phi) - \langle 2|\Delta\rangle \sin(2\pi\arg 1)\sin(2\pi\Phi)\big] \\ &+ 2\rangle \cdot a \cdot \big[\langle 1|\Sigma\rangle \cos(2\pi\arg 1)\cos(2\pi\Phi) - \langle 1|\Delta\rangle \sin(2\pi\arg 1)\sin(2\pi\Phi)\big]\end{aligned} \quad (4.75)$$

Again modifying the previous formulas, the two ASM received signals are

$$\begin{aligned}\Sigma = A_J(R) e^{\left[2\pi i\left\{(\beta\Phi + f_D T)p - \frac{2R_J}{\lambda}\right\}\right]} \Big[&\cos^2(2\pi\Phi)\{z \cdot \langle\Sigma|1\rangle\langle 2|\Sigma\rangle + \langle\Sigma|2\rangle\langle 1|\Sigma\rangle\} \\ &- \sin^2(2\pi\Phi)\{z \cdot \langle\Delta|1\rangle\langle 2|\Delta\rangle + \langle\Delta|2\rangle\langle 1|\Delta\rangle\} \\ &+ i\cos(2\pi\Phi)\sin(2\pi\Phi) \\ &(z \cdot \langle\Delta|1\rangle\langle 2|\Sigma\rangle + z \cdot \langle\Sigma|1\rangle\langle 2|\Delta\rangle + \langle\Delta|2\rangle\langle 1|\Sigma\rangle + \langle\Sigma|2\rangle\langle 1|\Delta\rangle)\Big]\end{aligned}$$
(4.76)

$$\begin{aligned}\Delta = A_J(R) e^{\left[2\pi i\left\{(\beta\Phi + f_D T)p - \frac{2R_J}{\lambda}\right\}\right]} \Big[&\cos^2(2\pi\Phi)\{z \cdot \langle\Delta|1\rangle\langle 2|\Sigma\rangle + \langle\Delta|2\rangle\langle 1|\Sigma\rangle\} \\ &- \sin^2(2\pi\Phi)\{z \cdot \langle\Sigma|1\rangle\langle 2|\Delta\rangle + \langle\Sigma|2\rangle\langle 1|\Delta\rangle\} \\ &+ i\cos(2\pi\Phi)\sin(2\pi\Phi) \\ &(z \cdot \langle\Sigma|1\rangle\langle 2|\Sigma\rangle + z \cdot \langle\Delta|1\rangle\langle 2|\Delta\rangle + \langle\Sigma|2\rangle\langle 1|\Sigma\rangle + \langle\Delta|2\rangle\langle 1|\Delta\rangle)\Big]\end{aligned}$$
(4.77)

These equations can be adapted to noise jamming or false-target generation via the phase of the exponential term, and they can be adapted to the various DCS configurations. Some common configurations are listed in Table 4.5.

Table 4.5
DCS EA Configurations

DCS	EA Antennas
Cross-polarization	Colocated; orthogonal polarizations
Cross-eye	Offset angles; identical polarizations
Double-cross	Offset angles; orthogonal polarizations

To gain insight into the EA, assume only the dominant terms for false-target generation. For simplicity, assume that the various antenna gains are real. To generate a false target at zero Doppler

$$\Sigma = A_J(R) e^{\left[2\pi i \left(\beta \Phi p - \frac{2R_J}{\lambda}\right)\right]} \cdot \cos^2(2\pi\Phi)(z+1) \cdot \langle 1|\Sigma\rangle\langle 2|\Sigma\rangle \tag{4.78}$$

$$\Delta = A_J(R) e^{\left[2\pi i \left(\beta \Phi p - \frac{2R_J}{\lambda}\right)\right]} \Big[\cos^2(2\pi\Phi)\{z \cdot \langle 1|\Delta\rangle\langle 2|\Sigma\rangle + \langle 2|\Delta\rangle\langle 1|\Sigma\rangle\} \\ + i\cos(2\pi\Phi)\sin(2\pi\Phi)(\{z+1\} \cdot \langle 1|\Sigma\rangle\langle 2|\Sigma\rangle)\Big] \tag{4.79}$$

After some algebra the monopulse ratio is

$$\frac{\Delta}{\Sigma} = W + iV = W_0 + iV_0 + \frac{1-z}{1+z} \cdot (\alpha + i\beta) \tag{4.80}$$

$$W_0 + iV_0 = i\tan(2\pi\Phi) + 0.5 \cdot \left[\frac{\langle 1|\Delta\rangle}{\langle 1|\Sigma\rangle} + \frac{\langle 2|\Delta\rangle}{\langle 2|\Sigma\rangle}\right] \tag{4.81}$$

$$-0.5 \cdot (\alpha + i\beta) = 0.5 \cdot \left[\frac{\langle 1|\Delta\rangle}{\langle 1|\Sigma\rangle} - \frac{\langle 2|\Delta\rangle}{\langle 2|\Sigma\rangle}\right] \tag{4.82}$$

For the sample case of phase monopulse, the terms of interest correspond to the imaginary part of the ratio or the value of V. Consider this value and the magnitude of the Σ channel signal from (4.78). Shifting the coordinate system in the space of the complex, monopulse control parameter gives the following results:

$$z = -1 + x + iy \tag{4.83}$$

$$V = V_0 - \frac{\beta}{2} + \frac{\beta x - \alpha y}{x^2 + y^2} \tag{4.84}$$

$$|\Sigma| = A_J(R) \cdot |\langle 1|\Sigma\rangle\langle 2|\Sigma\rangle| \cdot \cos^2(2\pi\Phi) \cdot \sqrt{(x^2 + y^2)} \tag{4.85}$$

$$P_\Sigma \approx a^2 \cdot \left[|z^2| + 2 \cdot \mathrm{Re}(z) + 1\right] \tag{4.86}$$

It is useful to examine the monopulse response in the z-plane to gain some intuitive understanding of DCS jamming. Consider first the Σ channel power. The first observation is that constant power corresponds to circles in the z-plane if a (or A) is fixed. If the outputs from both the ports are less than the maximum, constant amplitudes in the Σ channel are circles centered on

$$z = -1 \qquad (4.87)$$

Generally, the EA signal needs to be maximized by maximizing the output power from either transceiver 1 or 2. From the definitions, if transceiver 2 is maximized, then the magnitude of z must be less than or equal to 1. This corresponds to inside the unit circle in the z-plane. These circles are centered on $z = -1$.

Similarly, if transceiver 1 is maximized, then the product az is limited. This corresponds to the region outside of the unit circle in the z-plane. The loci in this case are larger radius circles centered at various points on the real axis. Figure 4.32 qualitatively indicates the results. It is important to note that the Σ-channel power is the maximum at $z = +1$ and the minimum (zero) at the complex pole of the ratio at $z = -1$. It is important to note that there is a region around this pole where the total power is insufficient to evoke a sensor response, because the signal level is too low and will be rejected by simple ASM system EP.

With this understanding, it is now useful to examine the monopulse ratio response in the z-plane. Define C as

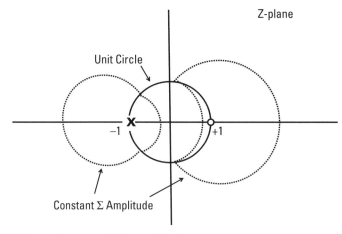

Figure 4.32 Loci of constant Σ-channel level.

$$C = V - V_0 + \frac{\beta}{2} \qquad (4.88)$$

The ratio of interest is then

$$C = \frac{\beta x - \alpha y}{x^2 + y^2} \qquad (4.89)$$

This expression has two possible solutions for values of C

$$y = \frac{\beta}{\alpha} \cdot x \qquad C = 0 \qquad (4.90)$$

$$\left(x - \frac{\beta}{2C}\right)^2 + \left(y - \frac{\alpha}{2C}\right)^2 = \frac{\alpha^2 + \beta^2}{4C^2} \qquad C \neq 0 \qquad (4.91)$$

First, examine the expression defining the parameter C. The primary interest is in the sensor response to EA in estimating the target angle relative to bore sight via the parameter V. As the EA is varied via jamming controls x and y, V is altered. The normal monopulse response (with no jamming) to a target at this angle is V_0. For small angles, V_0 is the standard measurement S-curve as shown in Figure 4.33. This occurs at $C = \beta/2$. If C increases then

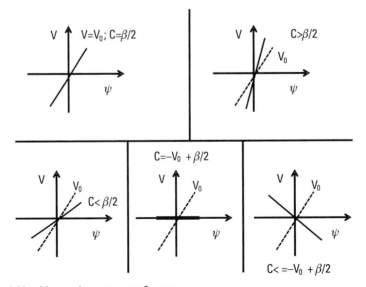

Figure 4.33 Monopulse response S-curve.

the monopulse response is stronger, but has the same sense. As C decreases, the response (V) eventually becomes zero and then reverses sense. When the curve reverses sense, the monopulse measurement relates that the target is in the opposite direction from bore sight from its true direction. These cases are illustrated in Figure 4.33.

Now the structure of the monopulse ratio response function in the z-plane can be illustrated. Figure 4.34 shows the key features for arbitrary values of α and β. The actual values depend on specifics of the EA system antennas and geometry.

In this figure, regions of interest are from the previous figure showing loci of constant amplitude in the Σ channel. Close to the monopulse ratio singularity, there is a region with insufficient power to be accepted by the sensor processing. (The maximum power is at $z = +1$.) As drawn, there are regions (shaded) where the monopulse response has been reversed (null rings). Eventually, these regions become correct sensing (labeled beacon rings). There are beacon rings on the opposite side of the singularity. The beacon to the right of the singularity and the repulsion to the left are the largest monopulse ratio values.

For example, Figure 4.35 illustrates the behavior of the monopulse measurement along the dashed straight line in the figure. Far from the singularity, the response is the normal response for a target at a particular angle off the antenna bore sight. Coming from the right-hand side, the magnitude of the

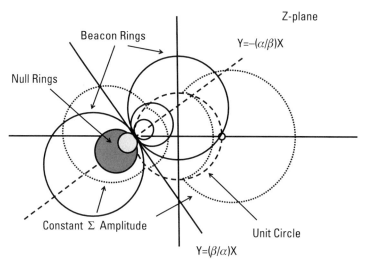

Figure 4.34 Monopulse response ratio in the z-plane.

ratio maintains its sense and increases. Just before the singularity the magnitude is very large (strong beacon). On the opposite side of the singularity the ratio is very large, but of the opposite sense (sign) indicating very strong induced angle error. Recall, however, that in the vicinity of the singularity the magnitude of the return in the Σ channel is very small. Continuing to the left, the sense of the monopulse ratio changes sign and again has the correct sense (beaconing response).

This understanding of DCS EA can assist in the task of interest. There are several DCS techniques to induce artificially large variance in the ASM target angle estimate. One technique is to set controls far enough from the singularity to register a measurement and then oscillate between beacon (just outside of the null rings) and repulsion (inside null rings). Another method would be to generate a pattern of z that circles the pole in the z-plane. A small enough circle ensures variations between strong beacon and strong repulsion.

For the illustration thus far, a retrodirective cross-polarization DCS EA system is considered. It has been assumed that the polarizations of the two EA source antennas are orthogonal. By definition

$$\langle 1|2 \rangle = g_1^{p*} \cdot g_2^{p} + g_1^{n*} \cdot g_2^{n} = 0 \qquad (4.92)$$

An often employed alternative method is to set the single EA antenna near to a cross-polarization state of the ASM antenna polarization (assumed known) and to mechanically oscillate the antenna by a small angular amount. This

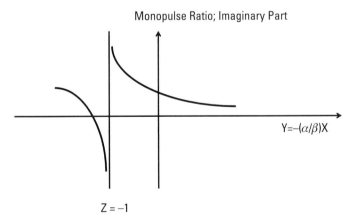

Figure 4.35 Example of monopulse ratio along a line in the z-plane.

hopefully has the effect of moving the EA response between null ring regions and beacon response regions.

Classically, it has been claimed that particular antennas are not vulnerable to cross-polarization EA. Several comments relative to cross-polarization EA are therefore appropriate. Any antenna viewed through an aerodynamic dielectric radome has polarization patterns near bore sight that resemble (qualitatively and quantitatively) the patterns of a standard parabolic antenna [15–18]. This statement is true for both transmit and receive patterns via reciprocity. These patterns are qualitatively illustrated in Figure 4.36. The plus and minus signs indicate the phase polarity relative to the phase of the Σ pattern on bore sight.

Thus, the pointing direction of the antenna can be manipulated via this vulnerability to cross-polarization EA. Rapidly changing from beacon to opposite sense (repulsion) can generate an artificially large variance of the ASM measure of monopulse ratio.

In addition to generating enhanced variance of the monopulse estimate, the hypothesis that the ASM sensor is tracking a false target can be measured via a modified jog detection (at the EA system) synchronized with the EA. This approach to probing the ASM sensor was tested during an Advanced Technology Demonstration in 1990. For example, mechanically oscillating the EA antenna at 5 Hz, if successful in deceiving the ASM system should induce a 5-Hz oscillation in the ASM antenna if the ASM is tracking the false target. (Of course, this oscillation may be mitigated by the antenna servo time constant for a sluggish servo system.)

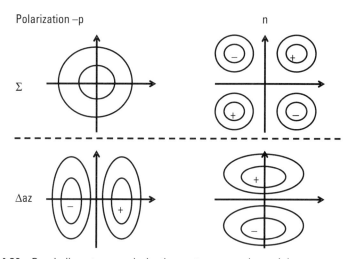

Figure 4.36 Parabolic antenna polarization patterns near bore sight.

This oscillation is a manifestation of the larger variance of the ASM monopulse measurements, and it is a means to probe the ASM system for EA effectiveness monitoring. The task for the EA system is to measure the effects of this oscillation via amplitude and/or polarization measurements. Figure 4.37 illustrates the polarization ratio of the ASM antenna (as viewed through the radome) along a typical azimuth of the ASM antenna.

To summarize if a false target is generated to seduce the track gates from the target, the BFD technique alerts the sensor processor to this possibility. At this point, the nonphysical pull on the tracking gates can be blocked. If false targets are generated (active or passive) they must have realistic RCS magnitude (or mean). Also, the RCS measurements must decorrelate from CPI to CPI or the false target can be distinguished from a real target by a rapid estimate of the Lag-1 value of the autocorrelation function.

The statistics of the monopulse measurement from a real target are distinguishable from those of an EA generated false target. A common technique is to measure the real part of the monopulse ratio. Another technique is to estimate the variance of the monopulse measurement. At shorter ranges (less than 10 to 12 nmi), the monopulse variance for a false target must increase to adequately mimic the statistics of an extended target. Several means exist to mimic these statistics. However, since the ASM sensor only accepts measurements from sizable targets (proper RCS), these EA techniques may be distinguished as well since the angular estimate statistic is strongly synchronized with RCS statistics.

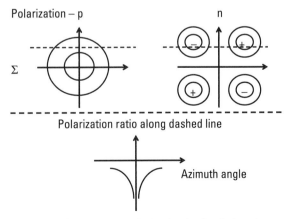

Figure 4.37 Example of ASM antenna polarization for fixed elevation.

If multiple false targets are generated via a DRFM-based EA, the various parameters of the targets must not be correlated with each other. And finally, a coarse (in range and Doppler resolutions) pulse Doppler radar seeker can adequately detect enough structure of the HVU in range and Doppler to distinguish the HVU from other ships as well as from chaff.

References

[1] Wehner, D., *High-Resolution Radar*, Boston, MA: Artech House, 1994.

[2] Tsui, J., *Digital Techniques for Wideband Receivers*, Norwood, MA: Artech House, 2001.

[3] Fouts, D., et al., "Single-Chip False Target Radar Image Generator for Countering Wideband Imaging Radars," *IEEE Journal of Solid-State Circuits*, Vol. 37, No. 6, June 2002, pp. 751–759.

[4] Morris, G., and L. Harkness, *Airborne Pulsed Doppler Radar*, Boston, MA: Artech House, 1996.

[5] Skolnik, M., *Radar Handbook*, Boston, MA: McGraw Hill, 1990.

[6] Richards, M., *Fundamentals of Radar Signal Processing*, Boston, MA: McGraw Hill, 2005.

[7] Chang, Y., L. Shi, X. Wang, S. Pingxiao, *Advanced Polarization Estimation Method using Spatial Polarization Characteristics of Antenna*, 2013 European Microwave Conference, pp. 1703-1706.

[8] Hongya, L., and J. Xin, *Methods to Recognize False Target Generated by Digital-Image Synthesizer*, International Symposium on Information Science and Engineering, Washington, DC: IEEE Computer Society, 2008, pp. 71–75.

[9] LongFei, S., W. XueSong, and X. ShunPing, "Polarization Discrimination Between Repeater False-Target and Radar Target," *Science in China Series F: information Sciences*, Vol. 52, No. 1, January 2009, pp. 149–158.

[10] Dai, H., Y. Chang, J. Li, "A New Polarization Estimation Method based on Spatial Polarization Characteristics of Antenna," *IEICE Electronics Express*, Vol. 9, No. 10, May, 2012, pp. 902-907.

[11] C. Tao, et al., "Polarization Identification of Passive Radar Target," *Journal of Projectiles, Rockets, Missiles, and Guidance*, Issue 2, 2013, pp. 109–112.

[12] Wang, Z., C. Mo, and H. Dai, "Interference Signal Suppression by Polarization Filters under Estimate Error," *Procedia Computer Science*, Vol. 107, 2017, pp. 503–512.

[13] Zong, Z., et al., "Detection Discrimination Method for Multiple Repeater False Targets Based on Radar Polarization Echoes," *Radioengineering*, Vol. 23, No. 1, April 2014, pp. 104–110.

[14] Guangfu, T., et al., "A Novel Discrimination Method of Ship and Chaff Based on Sparseness for Naval Radar," *IEEE Conference on Radar Conference*, Rome, Italy, May 26–30, 2008, pp. 1–4.

[15] Sherman, S. M., and D. K. Barton, *Monopulse Principles and Techniques*, 2nd edition, Norwood, MA: Artech House, 2011.

[16] Schleher, D. C., *Electronic Warfare in the Information Age*, Norwood, MA: Artech House, 1999.

[17] MacGrath, D., "Analysis of Radome Induced Cross Polarization (U)," WL-TM-92-700-APN, USAF, Washington, DC, March 1992.

[18] Ostrovityanov, R., and F. Basalov, *Statistical Theory of Extended Radar Targets*, Norwood, MA: Artech House, 1985.

5
LPI Radar EP Waveforms

In this chapter, several EP-specific classes of ASM radar waveforms are described. The radar waveform is designed to provide the optimal estimates of needed guidance parameters with minimum vulnerability to EA. The modern ASM radar sensor is coherent pulsed Doppler radar utilizing LPI waveforms. The modern radar can readily detect targets and measure the required guidance parameters. The EW battle is an information battle focusing on the target classification task. Multiple target features are measured specifically for the purpose of reliable target classification. In the previous chapter, target features generally associated with the physical nature of the target were discussed as a means to distinguish deceptive targets from real targets. The waveforms described in this chapter are specifically designed to enhance the ASM capability to identify the correct target when in an environment with deliberate EA.

In the standard hardware configuration for LPI radar, first the ASM radar receivers collect digital samples. Then the sensor performs coherent pulse compression and finally coherent Doppler processing. These data samples are then processed with rapid DSP algorithms. The modern ASM radar makes use of advanced technological microwave components that make possible the development of sophisticated waveforms that provide the necessary guidance information while deliberately making the correct target more readily identifiable in the EA environment.

In the first section of this chapter, several aspects of the waveform pulse compression codes are examined for the purpose of enhanced target classification. An obvious technique is to randomize the codes from one pulse to another within a single coherent processing interval. This makes it impossible to generate proper false targets at ranges less than that of the EA platform or even at longer ranges that are close to the platform range based on the code length. This makes it impossible to generate adequate false target EA that competes with the ship echo as a viable target from onboard that ship. Rather the EA must be placed on a platform at a much shorter range in the direction of the threat ASM. This requires an attack alert significantly prior to the attack and also the means to communicate with the platform. The special case of a LFM waveform is examined. In this case, a particular means of generating a target in front of the EA platform is examined that is based on understanding the ASM DSP.

In the second section, a stepped frequency waveform is examined. This waveform is utilized to deliberately enhance the characteristics of an extended target in contrast to the features of a false target. This makes clear the concept of deliberately generating waveforms for the purpose of probing the target for classification information via the EW signal processing. This technique is developed by People's Republic of China (PRC) engineers in a series of publications. Also, a variation of this technique is detailed in a patent granted to an ASM engineer from the Soviet Union in the 1990s.

Exploiting the modern microwave components and the sophisticated signal processing, it is now feasible to generate multiple variations of transmit waveforms during the same CPI to further confound the EA system. This technique is discussed relative to a combination of sequential waveforms into a single waveform that can be processed in a variety of ways based on the real-time progression of the attack scenario.

In the final section, the impact of ASM motions generated ostensibly to confound kinetic defensive weapons fire control systems is shown to be useful for EP. The ASM terminal phase weave maneuver is shown to induce radar signal processing effects in a manner that again enhances the characteristics of an extended naval target. This is another example of deliberate actions for the purpose of countering EA and for the purpose of enhanced target classification.

5.1 Coded Waveforms EP

The pulse compression codes utilized by the various ASM radar sensors include many of the standard phase and frequency codes such as Barker codes and LFM or chirp. Pulse compression is used to provide a finer range resolution in the range swath of interest, while using a wider pulse in time (tens of microseconds) at low peak energy (hundreds of Watts). The typical range resolution of present ASM radar is 10 to 30m. A typical PW can be as short as 1 μs or as long as 300 μs. More than adequate detection of naval targets can thus be achieved with a peak power of several 100W or even less than 100W. The result of these techniques is sufficient range resolution and adequate total energy for detection with an LPI transmit signal that is difficult for the EA system to detect [1, 2].

As described in Chapter 2, consider a phase coded signal. Define a narrow fixed frequency pulse with a range resolution of 30m or

$$PW_n = 0.2 \ \mu sec \qquad (5.1)$$

The transmit signal is represented as [for time within $(-PW_n/2, PW_n/2)$]

$$s_{Tn}(t) = s_0 \cos\left[2\pi\left(f_T t + \omega_n\right)\right] \qquad (5.2)$$

The received signal (Σ or Δ) complex amplitude for pulse p of the CPI is (after radar processing and coherent digital sampling of the narrow pulse)

$$s_{Rn}(p) = s_1 \cdot e^{2\pi i \left(\beta \Phi + f_D T\right) p} \cdot e^{2\pi i \left(-\frac{2R_0}{\lambda}\right)} \cdot e^{2\pi i \omega_n} \qquad (5.3)$$

As before, the radar transmits N of these contiguous subpulses as a single long pulse of PW

$$PW = N \cdot PW_n \qquad (5.4)$$

From (2.73) and (2.81), the time samples (equivalently the range samples) are processed via the matched filter for pulse p developed from the known transmit pulse. (For illustrative purposes, the matched filter is centered on a nominal range cell or time sample.)

$$h_n(p) = s^*_{T(-n)}(p) = e^{-2\pi i \omega_{-n}} \qquad n = -\frac{N}{2} \ldots \frac{N}{2} \qquad (5.5)$$

$$\chi_m(p) = \sum s_{Rn}(p) \cdot h_{m-n}(p) \qquad (5.6)$$

The radar processor uses the known coded sequence of the phases (ω_n) to combine the N subpulses via the known coherent matched filter as

$$\chi_m(p) = \sum s_{Rn}(p) \cdot e^{-2\pi i \omega_{n-m}} \qquad (5.7)$$

The resulting peak value is (neglecting any straddle losses) approximately

$$\chi_0(p) = N \cdot s_1 \cdot e^{2\pi i (\beta\Phi + f_D)p} \cdot e^{2\pi i \left(-\frac{2R_0}{\lambda}\right)} \qquad (5.8)$$

After this range compression the complex sample has amplitude increased by the multiplicative factor N and the identical phase and range resolution of each of the subpulses. For each transmit pulse, samples are collected and pulse compression processing is implemented to generate the array of range samples for each pulse. The standard procedure is to next complete the Doppler matched filter processing of the array of range time data for the P pulses. After the last pulse of the CPI is processed in this manner the pulse compression outputs are Fourier transformed to develop the Doppler range array.

However, the range compression and the Doppler processing are independent linear processes. As long as coherency is maintained from pulse to pulse the Doppler result is not dependent on the particular phase shift key (PSK) code used. Suppose the ASM radar sensor controller contains the identical long PSK code with N chips ($N > 30$), but with each code starting at the subsequent chip of the original code. Thus, N different codes with identical gain response but perhaps varying side lobe structure are available. As a simple illustration, consider the seven codes beginning with the 7-chip Barker code as listed in Table 5.1.

Each of these codes has the exact same range gain response, leading to samples like the samples represented in (5.8). Suppose the radar sensor randomly chooses the phase code number for a particular pulse p of the CPI from this table. The full Doppler range processing in the primary cell will not be degraded or altered, except maybe the subtleties of the range side lobes.

The DRFM-based EA system may attempt to capture this LPI signal and repeat a single seduction false target. Or the EA system may repeat many false targets at various ranges and Doppler values in an attempt to confuse the ASM sensor target classification task. Assume that the ASM goal is to attack a target ship within a range swath of several kilometers. To achieve a

30m range resolution the ASM radar generates a phase coded pulse with 30 or more chips of duration 0.2 μs for a PW of greater than 15 μs. Thus, the ASM transmit PW corresponds to a range extent of over 2 km, and any particular pulse can use any one of over 30 random code sequences [3].

For the DRFM-based EA system to copy and then retransmit the correct code requires that the false target be over 2 km behind the EA platform. Even if the EA system knows the general character of the LPI codes, any closer false target generated with an arbitrary code number will probably use the incorrect code. The use of the incorrect code will result in the false target being severely degraded upon coherent processing in the radar processor.

Thus, while false targets can be generated at a much longer range than that of the EA platform, the EA platform will be the closest viable target. For the example described, all of the viable false targets must be greater than 1 nmi behind the EA platform.

As a simple illustration the codes from Table 5.1 were used. Figure 5.1 illustrates the true range filter response for a single pulse, assuming the radar transmits and receives code #1: the Barker Code. The figure shows the amplitude maximum in range cell 0 and several range filter side lobes about 17 dB lower at nearby range cells.

Next, it is assumed that the radar transmits a CPI sequence of just four pulses. The radar uses codes #2, #4, #6, and #3. The solid line in Figure 5.2 is the amplitude response at the target range and for the neighboring range cells for this CPI sequence. Again, the range filter side lobes are about 17 dB below the peak value. It is assumed that the EA system generates a false target using code #1 for all four pulses in the CPI. The range filter response of the false target as processed by the radar signal processor is shown as the dashed line in the figure. It is seen that for this simple example the entire false-target

Table 5.1
7-Chip Barker Code Variants

Code Number	Code Sequence
#1	+ + + − − + −
#2	+ + − − + − +
#3	+ − − + − + +
#4	− − + − + + +
#5	− + − + + + −
#6	+ − + + + − −
#7	− + + + − − +

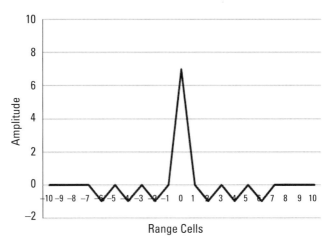

Figure 5.1 Simplistic filter response for 7-chip Barker code.

response is 17 dB below the true target response. (It is assumed that both the true target and the false target have 0-kts radial speed for this example.)

Mismatching the codes significantly degrades the total response. The result is more significant for a realistic sequence consisting of a longer range code (30+ chips) and a longer CPI (16+ pulses). This EP technique is to transmit a random time sequence of long LPI range coded pulses during the CPI.

Another aspect of false targets was discussed in the previous chapter. If the EA system attempts to generate a false target for a complicated pulse compression code sequence, it is expected that it will miss one or more of

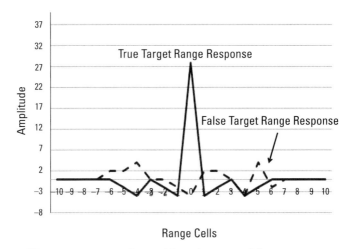

Figure 5.2 Filter response: 4 random 7-chip codes versus false target.

the pulses. If one or more pulses in the CPI are omitted in the CPI, multiple weak targets are generated at the same range, but at different Doppler filters. For example, a target at 0 kts would have identical phase for each pulse. If only every other pulse is received by the ASM radar, this is equivalent to two targets of differing phase history. Thus, when a target is detected the ASM signal processor inspects the returns at different Doppler filters at this particular range. The existence of several weak targets at the same range but with different radial speed (Doppler) is characteristic of false targets with missing pulses. This is illustrated in Figure 5.3.

Thus, when using a waveform with randomized pulse compression code the closest target in the range swath is typically the correct target of all the targets. In addition, Doppler ghost images for a single range cell are an indication of the EA system missing pulses, that is, false targets.

If the ASM radar employs an LFM code on the transmit pulses, the EA system may exploit this pulse compression code to generate short-range false targets. First, consider the LFM code for a static situation. The transmit pulse is

$$s_T(t) = s_0 \cos\left[2\pi\left(f_0 t + \frac{mt^2}{2}\right)\right] \qquad t \ni \left[\frac{-PW}{2}, \frac{PW}{2}\right] \qquad (5.9)$$

The parameter m relates the PW to the BW as

$$m = \frac{BW}{PW} \qquad (5.10)$$

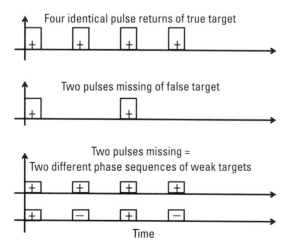

Figure 5.3 Sequence with missing pulses.

The instantaneous frequency of the modulation waveform is

$$f(t) = f_0 + mt \tag{5.11}$$

After complex processing in the receiver the match filter for this transmit signal is

$$\chi(\tau) = \int_{-PW/2}^{PW/2} dt \cdot x_R(t) \cdot x_T^*(t-\tau) \tag{5.12}$$

$$x_T^*(t-\tau) = e^{-2\pi i \cdot \frac{m(t-\tau)^2}{2}} = e^{-2\pi i \cdot \frac{m(t^2 - 2t\tau + \tau^2)}{2}} \tag{5.13}$$

The echo from a stationary object at range R_0 is

$$s_R(t) = s_1 \cos\left\{2\pi\left[f_0\left(t - \frac{2R_0}{c}\right) + \frac{m\left(t - \frac{2R_0}{c}\right)^2}{2}\right]\right\} \tag{5.14}$$

After processing in the radar receiver the complex received signal is

$$x_R(t) = s_1 \cdot e^{2\pi i \left[f_0\left(\frac{-2R_0}{c}\right) + m\left(\frac{2R_0^2}{c^2}\right)\right]} \cdot e^{2\pi i \left[m\left(\frac{-2R_0}{c}\right)t\right]} \cdot e^{2\pi i \left(\frac{mt^2}{2}\right)} \tag{5.15}$$

Using (5.12), (5.13), and (5.15), it is seen that the matched filter output is a sinc function. With (5.10), the result peaks at range R_0 and is

$$\chi(\tau) = K \cdot \text{sinc}\left[\pi BW \cdot \left(\tau - \frac{2R_0}{c}\right)\right] \tag{5.16}$$

In reality, the ASM platform is moving at high speed and the ship target may also be moving. In this work, it is assumed that the transmit frequency is offset to correct for the Doppler resulting from this ASM platform motion. Equation (5.9) becomes

$$s_T(t) = s_0 \cos\left\{2\pi\left[(f_0 + \delta f)t + \frac{mt^2}{2}\right]\right\} \tag{5.17}$$

$$\delta f = f_0 \cdot \frac{-2v\cos(\gamma)}{c} \quad (5.18)$$

The range to the target is now time-varying

$$R(t) = R_0 - v\cos(\varphi)t - v_T \cos(\theta_T)t \quad (5.19)$$

These geometry and dynamics variables were defined previously. The defining figure is repeated as Figure 5.4.

The echo signal is now

$$s_R(t) = s_1 \cos\left\{2\pi\left[(f_0 + \delta f)\left(t - \frac{2R(t)}{c}\right) + \frac{m\left(t - \frac{2R(t)}{c}\right)^2}{2}\right]\right\} \quad (5.20)$$

At this point, there are many terms that contribute to the match filter integration. Most of the terms do not depend on the integration variable t. There are several terms that are identical except for the scale f_0 or δf. Since f_0 is in gigahertz and δf is in megahertz or less, the corresponding δf terms can be neglected. Proceeding in this manner, the dominant terms that contribute to the integrand result in a correction to (5.16) for the dynamic case

$$\chi(\tau) = K \cdot \mathrm{sinc}\left[\pi \mathrm{BW} \cdot \left(\tau - \frac{2R_0}{c} + f_0 \frac{\mathrm{PW}}{\mathrm{BW}}\left\{\frac{2v}{c}(\cos\varphi - \cos\gamma) + \frac{2v_T}{c}\cos\theta_T\right\}\right)\right] \quad (5.21)$$

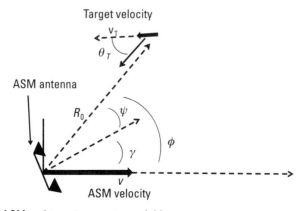

Figure 5.4 ASM and target geometry variables.

Thus the dynamics of the geometry introduces a range offset. This is a well-known property of the LFM ambiguity function [1–3]. This range error is of the order of

$$\Delta R \approx f_0 \cdot \frac{\text{PW}}{\text{BW}} \cdot v \cdot \sin\gamma \cdot \sin\psi \qquad (5.22)$$

Assume f_0 is about 10 GHz. Assuming the LFM PW of about 10 μs, a BW of tens of megahertz, and an ASM closing speed of Mach 3 gives a range error of the order of 0.01m.

At this point, consider the return from a false target generated by a DRFM-based EA system that deliberately adds a false Doppler shift of magnitude Δf to the false target using the standard narrowband approximation. The false target echo replaces (5.20) with

$$s_R(t) = s_1 \cos\left\{2\pi\left[(f_0 + \delta f + \Delta f)\left(t - \frac{2R(t)}{c}\right) + \frac{m\left(t - \frac{2R(t)}{c}\right)^2}{2}\right]\right\} \qquad (5.23)$$

This term again alters the output of the match filter. Neglecting the term from the dynamics illustrated in (5.21) and (5.22), the match filter output is approximately

$$\chi(\tau) = K \cdot \text{sinc}\left[\pi \text{BW} \cdot \left(\tau - \frac{2R_0}{c} + \Delta f \cdot \frac{\text{PW}}{\text{BW}}\right)\right] \qquad (5.24)$$

Using (2.111) and the result in (5.24), the measured range of this false target is now

$$\hat{R} = R_0 - \Delta f \cdot \frac{c \cdot \text{PW}}{2\text{BW}} = R_0 - \Delta f \cdot \text{PW} \cdot \delta R \qquad (5.25)$$

As an example assuming a range resolution of about 30m and a PW of 20 μs, the range expression for Δf in hertz gives

$$\hat{R} \approx R_0 - \Delta f \cdot 6 \times 10^{-4} \, \text{m} \qquad (5.26)$$

Thus, an offset frequency (Δf) of about 500 kHz added to the false target will make the false target appear to be about 300m in front (shorter range) of the platform, even though it is generated after reception in the EA system. The range estimate is the result of the corrupted range filtering performed by the ASM DSP. This is illustrated in Figure 5.5.

If the ASM seeker radar transmits an LFM coded LPI waveform, it is feasible for the DRFM-based EA system to generate a false target that artificially appears at a closer range than the platform by transmitting a signal with the properly offset frequency. Whether the false target is at a shorter range or longer range depends on the frequency shift sign being coordinated with the slope of the LFM (increasing or decreasing). To properly engage in the modern EW battle requires the EW engineer to fully understand the subtle features of the signal processing.

5.2 Stepped Waveforms EP

In Chapter 3 the mathematical model for a simple true target was developed, and the model for a simple false target was also developed. In Chapter 4, the model for an extended target consisting of many scatter elements was developed. It was shown that the RCS statistics of a true target can be distinguished from that of a simple false target. Also, the statistics of the monopulse ratio for a false target and an extended target were compared. During the previous discussion, it was shown that varying the carrier frequency is a standard radar technique for enhancing the RCS time variation of an extended or complex target.

Another well-known technique to enhance the distinguishing features of an extended target from the features of a false target is to utilize a stepped

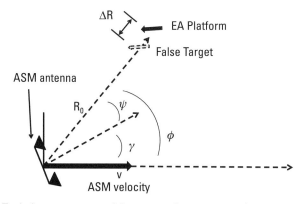

Figure 5.5 Technique to generate false target short-range estimate.

frequency waveform. Stepped frequency waveforms can be efficiently used to develop a range profile of a complex target [2, 4–6].

In addition, a stepped frequency waveform can be used as a target classification probe EP. In the previous derivation, a simple transmit waveform was utilized with transmit frequency from (3.62)

$$f_T = f_0 + \delta f \tag{5.27}$$

The following parameters were defined in Chapter 3 and are repeated here

$$\text{PRI} = T \tag{5.28}$$

$$\delta f = \frac{-2v\cos\gamma}{\lambda} \tag{5.29}$$

$$\beta = \frac{-4vT\sin\gamma}{\lambda} \tag{5.30}$$

$$f_D = \frac{2v_T \cos\theta_T}{\lambda} \tag{5.31}$$

$$\Phi = \frac{d}{2\lambda} \cdot \sin(\psi) \tag{5.32}$$

The variable in (5.29) is the offset of the transmit frequency from the receiver nominal carrier frequency used to correct for platform-induced Doppler along the antenna direction. The variable in (5.30) is a dimensionless variable representing ASM platform speed orthogonal to antenna pointing. The variable in (5.31) represents the target radial speed portion of the Doppler frequency. The variable in (5.32) represents the angle of the scatter element relative to the antenna bore sight.

The dominant target scatter element echo portion of the RF processing output at the analog to digital converter for pulse p for the Σ channel is

$$\Sigma_k = A_k(R) e^{\left[2\pi i \left\{ (\beta\Phi + f_D T) p - \frac{2R_k}{\lambda} \right\} \right]} \cos^2(2\pi\Phi) \tag{5.33}$$

This expression was used in Chapter 4 to show the result for the full target return as (4.63). The scatter element RCS is absorbed into the amplitude definition. While the RCS is random from one CPI to the next, it is generally confined to a single Doppler range cell. With this assumption, the dominant term is

$$\Sigma_T = e^{2\pi i(\beta\Phi + f_D T)p} \cdot e^{-2\pi i \frac{2R_0}{\lambda}} \cos^2(2\pi\Phi) \cdot \left(\sum a_k \cdot e^{2\pi i \frac{2\delta R_k}{\lambda}}\right) \quad (5.34)$$

Approximately

$$\Sigma_T = A_T \cdot e^{2\pi i(\beta\Phi + f_D T)p} \cdot e^{-2\pi i \frac{2R_0}{\lambda}} \cos^2(2\pi\Phi) \quad (5.35)$$

Similarly, the return from a false target (F) from a DRFM-based EA system or from a passive decoy is represented as

$$\Sigma_F = A_F \cdot e^{2\pi i(\beta\Phi + f_D T)p} \cdot e^{-2\pi i \frac{2R_0}{\lambda}} \cos^2(2\pi\Phi) \quad (5.36)$$

From the discussion in Chapter 2, the discrete Fourier transform Doppler processing applied to (5.35) or (5.36) results in a sinc function filter output (assuming no filter weighting function) for Doppler cell m

$$\Sigma = A \cdot e^{-2\pi i \frac{2R_0}{\lambda}} \cdot \cos^2(2\pi\Phi) \cdot \text{sinc}\left\{\pi\left[(\beta\Phi + f_D T)P - m\right]\right\} \quad (5.37)$$

The Doppler frequency and the frequency resolution are

$$F_D = \frac{\beta\Phi + f_D T}{T} \quad (5.38)$$

$$\Delta F_D = \frac{1}{PT} \quad (5.39)$$

Again, the Doppler resolution is the inverse of the total CPI integration time.
Returning to the previous derivation, assume that the transmit frequency for pulse p of a particular CPI of P pulses stepped in frequency and is

$$f_{Tp} = f_0 + \delta f + p \cdot \Delta f \quad (5.40)$$

Comparing the result from the prior derivation, the new result for a target scatter element k is

$$\Sigma_k = A_k(R)e^{\left[2\pi i\left\{\left(\beta\Phi + f_D T - \frac{2\Delta f \cdot R_k}{c}\right)p - \frac{2R_k}{\lambda}\right\}\right]}\cos^2(2\pi\Phi) \qquad (5.41)$$

Summing over the multiple scatter elements gives the final result. The time delay to the sample gives the nominal range to the standard range cell. In (5.33) and (5.34), the pulse dependent portion did not depend on the individual scatter element range. The result of the Doppler processing was a peak amplitude in a single (wide in range) range/Doppler cell representing the complex RCS of the target. The RCS measurement in this single cell was examined in Chapter 4.

With the stepped frequency, the Doppler filter cells at this range now represent the angle off bore sight, the true target Doppler, and information about the range profile of the target. At this point, it is customary to evaluate the size of the frequency step and the relation to the size of the target for the purpose of obtaining an accurate range profile. For the single scatter element, the result of Doppler processing gives the following equation corresponding to (5.37):

$$\Sigma = A \cdot e^{-2\pi i \frac{2R_0}{\lambda}} \cdot \cos^2(2\pi\Phi)$$
$$\cdot \operatorname{sinc}\left\{\pi\left[\left(\beta\Phi + f_D T - \frac{2\Delta f \cdot R_k}{c}\right)P - m\right]\right\} \qquad (5.42)$$

The frequency and resolution are as above in (5.38) and (5.39) with the additional term

$$F_D = \frac{\beta\Phi + f_D T - \frac{2\Delta f \cdot R_k}{c}}{T} \qquad (5.43)$$

The portion of this frequency corresponding to the scatter element range gives a range resolution

$$\Delta R = \frac{c}{2 * P * \Delta f} \qquad (5.44)$$

This is the same expression for range resolution as before with the BW corresponding to the frequency range of the stepped waveform

$$BW = P \cdot \Delta f \qquad (5.45)$$

Assume the step is 5 MHz and $P = 64$ for a total BW of 320 MHz. Then the range resolution of the Doppler filters is about 0.5m and the full coverage of the 64 cells is about 32m. This choice of parameters will cause the ship target range profile to suffer from significant aliasing, since the range is many kilometers. However, the result is a response for the true target in the several Doppler cells at this range. To summarize, the response for a true target with a fixed frequency waveform peaks in a single Doppler range cell while the response with a stepped waveform smears over many of the Doppler cells within this range cell.

However, the goal for the ASM sensor is not to achieve an accurate range profile, but rather to enhance the difference between a false (point) target and a true (extended) target. The expression for the receiver output for a false target is the same as (5.42)

$$\Sigma_F = A \cdot e^{-2\pi i \frac{2 R_F}{\lambda}} \cdot \cos^2(2\pi\Phi) \\ \cdot \text{sinc}\left\{\pi\left[\left(\beta\Phi + f_D T - \frac{2\Delta f \cdot R_F}{c}\right)P - m\right]\right\} \qquad (5.46)$$

The difference is that this is the full return for the false target coming entirely from a single range and angle. Proceeding as above, if the waveform is a fixed frequency, the response is in a single Doppler range cell. If the waveform is a stepped frequency the meaning of the Doppler range cells is confounded, but the response is still in a single Doppler range cell.

Figures 5.6 and 5.7 represent the results of a simple simulation. Assume the ASM radar uses a fixed frequency waveform shown at the top of Figure 5.6. (The figure shows four pulses for illustration purposes.) A simulation was run with a target consisting of four scatter elements and a simple false target of the same total RCS. Both the targets are at 0-kts radial velocity. Figure 5.7 illustrates the Doppler processing output for both the false (point) target and for the ship HVU target for the fixed frequency waveform (top row). It is seen that the outputs are identical.

It is now assumed that the ASM radar uses a simple stepped frequency waveform as illustrated at the bottom of Figure 5.6. (Again, only four pulses are shown.) For the simulation, the frequency step level is 5 MHz and the CPI consists of 16 pulses. Figure 5.7 shows the Doppler processing result for the two simple targets in the bottom row. The peak for the false (point) target is of the same magnitude as before but falls in a different Doppler cell as expected. While the targets have exactly the same total RCS, the true target consisting

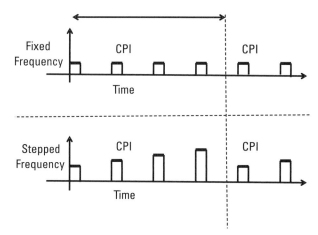

Figure 5.6 Alternating complex EP probe waveforms.

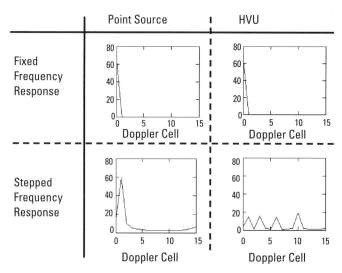

Figure 5.7 True target versus false target responses.

of four scatter elements at slightly differing ranges presents a response consisting of four individual peaks each of much less magnitude than the total RCS for the fixed frequency waveform.

As seen, the false target response occupies a single Doppler range cell with either waveform albeit at a different Doppler cell. However, if the true target is detected with the fixed waveform and then interrogated with the stepped waveform, the image smears over multiple cells representing an aliased high-resolution range profile. Thus, the waveform stimulates information to more clearly identify distinguishing features between the true targets and false targets. Many papers have been published by PRC personnel and others exploring these and related techniques [7–9].

Thus far the fixed frequency waveform and the stepped frequency waveform have been considered to be utilized in distinct CPIs. An interesting method can be proposed to further confound the EA system. With modern microwave technology, it is feasible to generate both transmit waveforms during the same CPI. As the transmit pulses are generated, a device such as an inexpensive direct digital synthesis can be used to create a fixed frequency pulse contiguous with a stepped frequency pulse sequence in the manner illustrated in Figure 5.8. (This should be compared with Figure 5.6.)

If the entire stepped sequence is at a frequency band distinct from the fixed frequency, the ASM radar receiver can be set to process either of the waveforms as before while transmitting both waveforms during the same CPI. This will further confound the EA system. Depending on the EA system processing, it is possible that the EA system will tune its system using an instantaneous frequency measurement (IFM) receiver. In this case, the EA system may only process the leading portion of the waveform, the fixed frequency portion in this illustration. If the EA system does not detect the stepped frequency portion, no false target with these properties will be generated. If the EA system does detect both portions of the waveform, the false target for both is still a point source.

As described, the ASM radar can process the leading portion of the waveform to detect and track the target for guidance inputs. By varying the

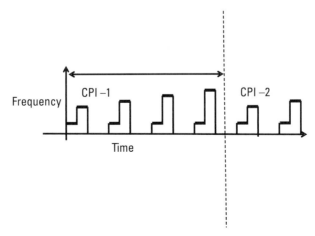

Figure 5.8 Complex simultaneous EP probe waveforms.

receiver parameters while still transmitting the same waveform the ASM radar can probe for target classification information with or without the EA system responding to the stepped frequency probe signal.

To summarize, the response for a true target with a fixed frequency waveform peaks in a single Doppler range cell while the response for a true target with a stepped waveform smears over many of the Doppler filters within this range cell. The ASM radar response for a false target to both waveforms is to peak in a single range Doppler cell. This is a simple means to probe the identity of a target. A possible counter would involve using a complex processor and a DRFM-based EA system to generate a much more realistic physical structure of a false target [10].

Several of the concepts discussed in this chapter and in Chapter 4 are somewhat combined in a patent granted in the 1990s to a Soviet engineer [5, 6]. The patent title being quite descriptive is *Method of Selecting Above-Water Targets*. The author discusses the need to distinguish among chaff, a passive decoy (e.g., a corner reflector), and a ship made up of many scatter elements when using a pulse Doppler coherent radar.

The proposed technique uses a fixed frequency waveform with Doppler processing (M cells) and an RCS threshold in the first stage. At this stage the author relates that the radar processor can identify chaff from its Doppler characteristics as described in Chapter 4. Once a potential target is detected, the proposed technique is to process a waveform to deliberately probe for additional target features. The author recognizes that shifting the waveform frequency by more than Δf enhances the rapid decorrelation of the RCS for ship targets (of length L), where

$$\Delta f > \frac{c}{2L} \quad (5.47)$$

The probe technique is to transmit N CPI waveforms at carrier frequencies (f_n) each increased by Δf or a stepped frequency waveform. For each CPI, M pulses are processed giving RCS estimates for CPI number n and Doppler cell m. For each CPI (n) the maximum RCS (σ_n) for the various M Doppler cells is selected, as qualitatively illustrated in Figure 5.9.

The author proposes that the normalized RCS metric be evaluated as

$$\text{Metric}' = \frac{\frac{1}{N}\sum_n |\sigma_{n+1} - \sigma_n|}{\frac{1}{N}\sum_n \sigma_n} \quad (5.48)$$

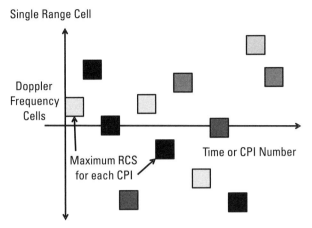

Figure 5.9 Maximum RCS estimate selected in each CPI.

Similar to the use of the Lag-1 estimator the author maintains that the RCS estimates decorrelate rapidly for ship targets, giving a value of the metric near one. In contrast, the author maintains that the passive (corner reflector) decoy does not decorrelate rapidly, giving a metric value near zero.

Thus, this early patent indicates the concept of exploiting Doppler range characteristics to identify chaff, exploiting RCS statistics for target classification, the concept of designing a waveform to probe for target features, and the concept of using a waveform that enhances target classification features. The direction of the technology is increasingly to develop waveforms and actions that deliberately probe the target for classification features.

5.3 Probe Waveforms EP

As a final means of exploiting the above result and probing the target for classification features, consider again the ASM Doppler measurement [7–9]

$$f_D = \frac{2f_0 v_T}{c} \cdot \cos\theta_T + \frac{2f_0 v}{c} \cdot (\cos\varphi - \cos\gamma) \tag{5.49}$$

For the moment, neglect the portion of the Doppler measurement that represents the target radial speed. The remaining portion of the Doppler measurement is a measure of the angle of any targets relative to the velocity vector direction. It is well known that the Doppler represents a grid of hyperbolas radiating outward from the ASM. The Doppler range array has the features of a two-dimensional grid of locations relative to the ASM velocity. (This same feature forms the basis of SAR imaging when looking to the side of the platform, orthogonal to the velocity.) The inherent Doppler measurement is symmetric about the ASM velocity. The ASM's own speed correction to the carrier frequency has the effect of setting the zero Doppler of this grid along the ASM antenna vector direction.

Suppose a fixed waveform is transmitted and the ASM trajectory is oscillated about the nominal direction by a small angle. For example, this is the typical ASM motion (weave trajectory) used by the ASM to mitigate the effectiveness of antimissile missile fire control solutions during the terminal phase of the engagement and is illustrated in Figure 5.10. The weave trajectory is generally in the direction of the target. If the ASM radar is tracking a target, the antenna remains generally pointed in the direction of the target during this maneuver. For simplicity, represent the endpoints of the extended target (B for bow and S for stern) as individual scatter elements.

Since the velocity vector is varying, the two-dimensional Doppler range coordinate grid is varying on the ocean surface relative to the target scatter elements. Several grid lines of fixed (angle) Doppler are illustrated in Figure 5.11.

The two scatter elements representing the ends of an extended target are at slightly differing azimuth angles for a broad-side attack. The Doppler value for each end is

$$\text{Doppler} f_k = f_0 \frac{2v}{c} \cdot \left[\cos\varphi_k - \cos\gamma\right] \quad (5.50)$$

The angle φ is the angle between the velocity vector and the target. The angle γ is the angle between the antenna and the velocity vector. Since the velocity vector is oscillating and the antenna is fixed in the direction pointing at the center of the ship target set

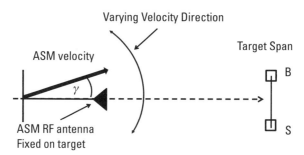

Figure 5.10 Tracking an extended target represented by two scatter elements.

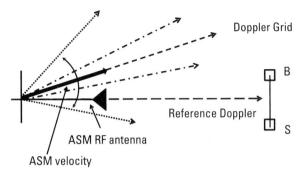

Figure 5.11 Doppler grid as a measure of angle from bore sight.

$$\varphi_k = \gamma \pm \delta\varphi_k \qquad (5.51)$$

$$\gamma = \gamma_0 \sin(\omega t) \qquad (5.52)$$

It is seen that the Doppler value for each end is different and this difference varies. Thus, the image of the extended target in the Doppler range array expands and contracts in the Doppler dimension during the engagement.

$$\text{Doppler } f_k B = f_0 \frac{v}{c} \cdot \left[\cos\gamma \left(\cos\delta\varphi_k - 1 \right) + \sin\gamma \cos\delta\varphi_k \right] \qquad (5.53)$$

$$\text{Doppler } f_k S = f_0 \frac{v}{c} \cdot \left[\cos\gamma \left(\cos\delta\varphi_k - 1 \right) - \sin\gamma \cos\delta\varphi_k \right] \qquad (5.54)$$

$$\text{Doppler extent} = 2 f_0 \frac{v}{c} \cdot \sin\gamma \cos\delta\varphi_k$$
$$\approx 2 f_0 \frac{v}{c} \cdot \gamma_0 \sin(\omega t) \cos\delta\varphi_k \qquad (5.55)$$

These varying Doppler values are synchronized with the ASM motion. (The result is analogous to Doppler beam sharpening.) This is another strong indication that the target is an extended object rather than a false target (point source). For the purpose of illustration, assume the ASM traveling toward a 200-m-long target at about Mach 3 and is performing 6-g turns as it approaches the target. A typical Doppler cell of about 30 Hz easily resolves the target. The left graph in Figure 5.12 displays the target endpoint scatter element Doppler values during a typical 3s of flight.

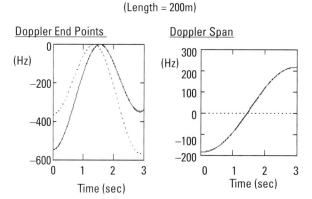

Figure 5.12 Doppler measurement of scatter elements.

The Doppler values are seen to vary over a 600-Hz range. The right graph displays the difference in these Doppler values as shown in (5.55). It is seen that the ship target is about 200 Hz long in the Doppler dimension of the Doppler range array. During the 3 sec of ASM flight shown, the ship target shrinks down to a single Doppler cell and again expands to about 200 Hz long. As indicated above, this behavior is similar to Doppler beam sharpening in the sense that the target length (including hotspots and structure) is resolved when the ASM is not flying directly at the ship target. The target is not resolved (in Doppler imaging) when the ASM velocity vector points in the direction of the target. All that is required for the target classification goal of the ASM is to observe this oscillating Doppler length behavior to indicate that the target is 200m long. A point target from a passive decoy or a DRFM-based EA system would remain collapsed into a single Doppler cell during the maneuver.

The tactical ASM maneuver performed for the purpose of mitigating kinetic weapon effectiveness provides a Doppler beam sharpening effect (EP) that periodically resolves the HVU length. This feature clearly distinguishes the HVU from the point image of a false target that does not vary in Doppler length with time.

In summary, the ASM sensor waveform design can include not only adequate target detection and accurate guidance parameter estimation with minimal probability of detection (LPI characteristics) by the EA system, but the waveforms can also be designed to enhance the classification features of the ship targets and false targets or to probe the targets for classification information. In this manner, while the ASM sensor detection and guidance performance is not degraded, the waveforms also serve to protect (EP) the ASM radar sensor from false target deception.

Several common and rapid approaches that are commonly used in modern ASM pulsed Doppler radar have been described in this chapter. These waveform design approaches include the following:

- Use of a long pulse compression code that is random from pulse to pulse;
- Detection of missed false echo pulses within the CPI;
- Alternating a fixed frequency waveform with a stepped frequency waveform;
- Commingling a fixed frequency waveform with a stepped frequency waveform;
- Examining the Doppler structure of the target during the ASM maneuver.

References

[1] Pace, P. E., *Detecting and Classifying Low Probability of Intercept Radar*, Norwood, MA: Artech House, 2009.

[2] Richards, M., *Fundamentals of Radar Signal Processing*, New York, NY: McGraw-Hill, 2005.

[3] Skolnik, M., *Radar Handbook*, Boston, MA: McGraw Hill, 1990.

[4] Schleher, D. C., *Electronic Warfare in the Information Age*, Norwood, MA: Artech House, 1999.

[5] Baskovich, E., et al., Method of Selecting Above-Water Targets, Patent RU 2083996, 1995.

[6] Ostrovityanov, R., and F. Basalov, *Statistical Theory of Extended Radar Targets*, Norwood, MA: Artech House, 1985.

[7] Hongya, L., et al., "Methods to Recognize False Target Generated by Digital-Image Synthesizer," *IEEE International Symposium on Information Science and Engineering*, Shanghai, China, December 20–22, 2008, pp. 71–75.

[8] Chen, H., et al., "A New Approach for Synthesizing the Range Profile of Moving Targets via Stepped-Frequency Waveforms," *IEEE Geoscience and Remote Sensing Letters*, Vol. 3, No. 3, July 2006, pp. 406–409.

[9] Shen, Y., et al., "A Step Pulse Train Design for High Resolution Range Imaging with Doppler Resolution Processing," *Chinese Journal of Electronics*, Vol. 8, No. 2, 1999, pp. 196–199.

[10] Fouts, D., et al., "Single-Chip False Target Radar Image Generator for Countering Wideband Imaging Radars," *IEEE Journal of Solid-State Circuits*, Vol. 37, No. 6, June 2002, pp. 751–759.

6

Multiple Receiver EP Signal Processing

This chapter introduces EP signal processing algorithms that more fully exploit the fact that there are multiple receivers in the sensor. The original angle estimation techniques using scanning antennas were vulnerable to simple angle deceptive EA that attacked the time processing required to measure the angle to the target. Classically, multiple sensor receivers were introduced for the purpose of improved monopulse processing, providing accurate and hardened estimation of the target angle on a single pulse through the simultaneous processing of multiple receivers.

Until now, all of the ASM tactical functions, such as logic decisions, detection, classification, tracking, and EP, are accomplished with the data array of the Σ channel. Using a separate receiver, the RF processing is fully duplicated in the Δ channel. The data of the Doppler range cell of interest in the Σ channel is combined with the data from the corresponding cell in the Δ channel array for monopulse measurement of the target angle. This measurement is input to the guidance subsystem for target tracking and ASM impact guidance.

However, with the introduction of modern microwave technology and high-speed digital processors, it is now feasible to introduce optimal multiple channel digital signal processing of the full array of data. This leads to immediate improved performance of several of the ASM functions since the ASM now utilizes all of the data available for all of its target measurements.

In addition, the use of optimal digital signal processing makes possible novel and efficient EP capabilities specifically effective against cover jamming. The goal of cover or noise jamming is to fill the ASM Σ channel Doppler range swath with high-level noise so that the HVU cannot be observed. In this way the ASM cannot isolate the correct cell in the array containing target information. At best the ASM can go into the HOJ mode.

While in HOJ the ASM can track the EA bearing only. A sea-skimming ASM may eventually hit the EA source or burn-through may finally uncover the desired target. If this EA is transmitted from a protection ship, then once the angle to the HVU has been sufficiently deceived relative to the HVU, alternative techniques are used to protect the jamming target platform. If the EA is generated by an off-board decoy or drone (e.g., unmanned surface vehicle), this EA is particularly effective in deceiving the ASM sensor.

In the first section, it is shown that information from the two radar receivers can be used to locate and track the hidden ship targets via a detailed examination of the monopulse ratio. This may require the extra processing burden of monopulse calculation for all or many of the Doppler range samples. However, by using the full array of monopulse ratios, it is possible to significantly mitigate cover jamming.

In the next section, optimal signal processing using both the receivers is examined. For the example of EA false target deception, it is shown how to extend the customary Σ channel processing to include the full data array. The prior results of EP algorithms are shown to be further improved by including the full two-channel array of data. Performance is at least 3 dB improved as a result of extending the algorithms to full array optimal processing.

In the final section, it is shown that optimal two-channel processing generally and significantly mitigates the effectiveness of cover jamming EA. Instead of having to perform a full array of complex division, straightforward linear algebraic techniques are employed to optimally process the full array of data.

The development of these signal processing algorithms has been pioneered by People's Republic of China (PRC) personnel in a series of published research papers on the subject of the degradation of space time adaptive processing (STAP) when applied to two-channel radar viewing in the more forward direction of the platform motion. It is shown herein that their research is directly applicable to the case of ASM radar sensor processing for the purpose of mitigating naval EA cover jamming. It is shown that this EP technique makes it possible to simplify the ASM hardware architecture, and the performance of the EP algorithm improves as the cover jamming EA level increases.

6.1 Dual-Coherent Source EP Approximation

The discussion in this section relates to an approximation to optimal, multiple channel processing for the purpose of mitigating noise or cover EA. By examining this algorithm, many of the later concepts can be better and more intuitively understood. Figure 6.1 illustrates the typical processing scheme presently employed by the ASM two-channel pulsed Doppler radar seeker.

Whenever a pulse is transmitted, possible echo information from a particular range swath is processed. The RF data is captured via the antenna subarrays. The RF data in the antenna subarrays is combined via the antenna hybrid to form the Σ and Δ beam patterns for later monopulse processing. These signals are processed separately in an identical manner via the two parallel RF receivers resulting in two complex arrays of digital signals. As described above, each of these arrays are processed if necessary via pulse compression to result in two arrays of range data for each pulse. Each of these arrays is again processed for a CPI over P pulses to generate two Doppler range arrays of digital data. The resultant Σ digital Doppler range array of data is analyzed via DSP in the array of digital processors indicated by the DSP box for target detection and parameter extraction. Logical decisions for seeker mode selection, target classification, target tracking, EP, automatic gain control (AGC) algorithms, and so on are all made based on Σ channel data in the post processing digital processor. The chosen cell in the Σ array for the possible target of interest is utilized to identify corresponding Doppler range cell in the Δ channel. These two complex data values are combined as the monopulse ratio for angle estimation and guidance input.

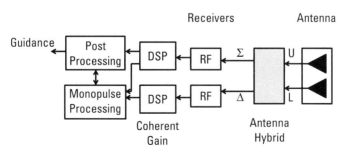

Figure 6.1 Standard ASM radar processing scheme.

For a standard pulsed Doppler radar, it is customary to view the Σ channel data as a Doppler range array as shown in Figure 6.2. In this figure, the HVU echo and the echo from the EA platform are indicated as not visible through the cover jamming. The EA cover jamming is indicated by the noise shading in all Doppler range cells of the array.

In reality, the data is a Doppler range angle array since the data contains information about the echo source angle relative to antenna Σ beam bore sight as evidenced by the capability to measure this angle via the monopulse ratio. For convenience at this time and as in the standard STAP technique, the data can again be viewed in two dimensions but as a range angle array.

Figure 6.3 qualitatively represents the cover jamming EA filling the array with the HVU and EA source deeply buried in the EA jamming. The shading is meant to heuristically represent complete cover jamming in all data cells with the Σ channel antenna beam pattern qualitatively weighting the jamming level. It is assumed that the antenna is pointing at the EA source when the ASM system is in HOJ mode.

Consider the Doppler range angle array of data for the full array of Σ channel and Δ channel. For every CPI and for each range cell, there are P Σ channel complex Doppler data values and P Δ channel Doppler values. In the presence of cover EA, the following hypothesis test can be performed at each Σ channel array cell. The Δ channel data is included for completeness. (The EA source is not of interest and is not explicitly shown in the equations for now.)

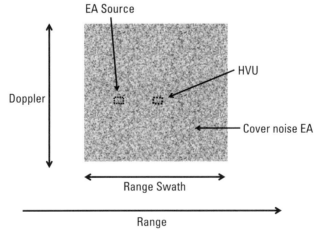

Figure 6.2 Doppler range array of data in presence of cover jamming.

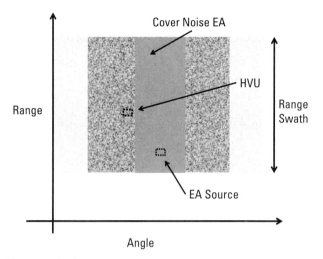

Figure 6.3 Range angle data array.

$$H0: \Sigma(H0) = PA_S\Sigma_S + YA_J\Sigma_J + \sigma_N\Sigma_N \quad (6.1)$$

$$\Delta(H0) = PA_S\Delta_S + YA_J\Delta_J + \sigma_N\Delta_N \quad (6.2)$$

$$H1: \Sigma(H1) = YA_J\Sigma_J + \sigma_N\Sigma_N \quad (6.3)$$

$$\Delta(H1) = YA_J\Delta_J + \sigma_N\Delta_N \quad (6.4)$$

In these formulas, the A_S and A_J represent the amplitude of the ship target and the jamming level, respectively. The σ_N represents the receiver noise level. The other symbols are a shorthand for the full quantities from the results of the previous chapters. From (3.83) through (3.87), the expression for Σ_S and the expression for Δ_S can be represented as

$$\Sigma_S = e^{\left[-2\pi i \frac{2R_S}{\lambda}\right]} \cdot \cos^2(2\pi\Phi)\langle\Sigma|\Omega|\Sigma\rangle \quad (6.5)$$

$$\Delta_S = e^{\left[-2\pi i \frac{2R_S}{\lambda}\right]} \cdot i\cos(2\pi\Phi)\sin(2\pi\Phi)\langle\Sigma|\Omega|\Sigma\rangle \quad (6.6)$$

$$\langle\Sigma|\Omega|\Sigma\rangle \approx \left|g_\Sigma^p(\psi)\right|^2 \sigma_{pp} \quad (6.7)$$

For convenience in this section, absorb the radar cross section and antenna terms into the amplitude. Then using the small angle approximations

$$\Sigma_S \approx e^{\left[-2\pi i \frac{2R_S}{\lambda}\right]} \tag{6.8}$$

$$\Delta_S = \Sigma_S \cdot i 2\pi \Phi_S \tag{6.9}$$

Similarly from (3.100) through (3.103), the expression for Σ_J and the expression for Δ_J can be represented as

$$\Sigma_J = e^{\left[2\pi i\left(-\frac{2R_J}{\lambda}\right)\right]} \left[\cos^2(2\pi\Phi)\langle\Sigma|\Sigma_J\rangle\langle\Sigma_J|\Sigma\rangle\right] \tag{6.10}$$

$$\Delta_J = e^{\left[2\pi i\left(-\frac{2R_J}{\lambda}\right)\right]} \left[i\cos(2\pi\Phi)\sin(2\pi\Phi)\left(\langle\Sigma|\Sigma_J\rangle\langle\Sigma_J|\Sigma\rangle\right)\right] \tag{6.11}$$

Implementing the same approximations for the EA source as for the target

$$\Sigma_J = e^{\left[2\pi i\left(-\frac{2R_J}{\lambda}\right)\right]} \tag{6.12}$$

$$\Delta_J = \Sigma_J \cdot i 2\pi \Phi_J \tag{6.13}$$

Define the variable Y as the result of the Doppler processing of the EA portion of the return in a particular Doppler cell. For later use the parameter Y is

$$Y(k) = \sum_p e^{-2\pi i \frac{kp}{P}} \cdot e^{\left[2\pi i\left(\beta\Phi p + \eta'_p\right)\right]} \tag{6.14}$$

The variables for Σ_N and Δ_N represent random white noise variables that are not correlated with each other. The expected value of each variable with itself is taken to be 1 and all other correlations of the noise variables are assumed to be 0.

Assuming P pulses, it follows from Chapter 3 that the target SNR in the Σ array is

$$\text{SNR} = \frac{PA_S^2}{\sigma_N^2} \tag{6.15}$$

Similarly, using the definition of Y in (6.14), the signal power to jamming power ratio (SJR) is

$$\text{SJR} = \frac{PA_S^2}{A_J^2} \qquad (6.16)$$

For completeness the full signal to interference level (SIR) is

$$\text{SIR} = \frac{PA_S^2}{A_J^2 + \sigma_N^2} \qquad (6.17)$$

Since the EA source and the target ship are both assumed to be within the main beam of the antenna, the hypothesis test can be performed with these small angle approximations in the above expressions. If H0 is true (for the Doppler range cell containing the ship echo)

$$\text{H0:}\ \Delta(\text{H0}) = i2\pi PA_S \Phi_S \Sigma_S + i2\pi Y A_J \Phi_J \Sigma_J + \sigma_N \Delta_N \qquad (6.18)$$

Similarly, if hypothesis H1 is true (for any Doppler range data cell not containing the ship echo), then the A_S terms are zero and

$$\text{H1:}\ \Delta(\text{H1}) = i2\pi Y A_J \Phi_J \Sigma_J + \sigma_N \Delta_N \qquad (6.19)$$

Suppose the monopulse ratio and scaling is performed for each cell. Assuming the jamming power is much larger than the receiver noise power, the general value is

$$k\frac{\Delta}{\Sigma} \approx i\psi_J + i\delta\psi \cdot \frac{z}{1+z} + \text{nse} \qquad (6.20)$$

$$z \equiv \frac{PA_S \Sigma_S}{YA_J \Sigma_J} \qquad (6.21)$$

$$\left\langle (|z|)^2 \right\rangle \approx \text{SJR} \qquad (6.22)$$

$$\delta\psi = \psi_S - \psi_J \qquad (6.23)$$

The complex variable z contains the processing gain as well as the radar range equation terms. The variable z is 0 for all cells (H1 true) except when the cell under test (CUT) contains the ship return (H0 true). The magnitude of the variable z is proportional to the square root of the ratio of the ship target power to the EA system jamming power.

The noise term in (6.20) is the standard monopulse measurement noise. The monopulse measurement noise variance is proportional to the inverse ratio of EA system jamming power to the receiver noise power (jam to noise power ratio [JNR]). A standard expression for this term will be detailed below.

Thus, after standard monopulse processing of the CPI digital data arrays the hypothesis test can be performed on the Doppler range array of monopulse ratios. For example, one could form a CFAR-like mask and compare the CUT with the background monopulse ratio level. The hypothesis test for each cell is then

$$\text{H0: } k\frac{\Delta}{\Sigma}(\text{H0}) \approx i\psi_J + i\delta\psi \cdot \frac{z}{1+z} + \text{nse} \qquad (6.24)$$

$$\text{H1: } k\frac{\Delta}{\Sigma}(\text{H1}) \approx i\psi_J + \text{nse} \qquad (6.25)$$

The monopulse ratio in every cell of the array is a variable from a probability distribution with identical noise statistics, but a different mean. This is a simple and well-known statistical hypothesis testing problem.

Consider the hypothesis test using a simple log likelihood ratio (LLR) model. Assume the measurement Z is a mean plus Gaussian noise with variance σ^2

$$\text{H0: } Z = \mu_0 + \text{nse} \qquad (6.26)$$

$$\text{H1: } Z = \mu_1 + \text{nse} \qquad (6.27)$$

LLR is the logarithm of the ratio of the two Gaussian probability distribution functions (for $i = 0$ or $i = 1$)

$$P(Z|Hi) = Ke^{-\frac{(Z-\mu_i)^2}{2\sigma^2}} \qquad (6.28)$$

$$\lambda = \ln\left[\frac{P(Z|\text{H0})}{P(Z|\text{H1})}\right] \qquad (6.29)$$

For this simple case

$$\lambda = \left(Z - \frac{\mu_0 + \mu_1}{2} \right) \cdot \frac{\mu_0 - \mu_1}{\sigma^2} \qquad (6.30)$$

The LLR variable (λ) is a random variable with Gaussian probability distribution. The hypothesis test is controlled by the variable Λ, where

$$\langle \lambda_0 \rangle = \frac{\Lambda}{2} \qquad (6.31)$$

$$\langle \lambda_1 \rangle = \frac{-\Lambda}{2} \qquad (6.32)$$

$$\text{Variance}(\lambda) = \Lambda = \frac{(\mu_0 - \mu_1)^2}{\sigma^2} \qquad (6.33)$$

The LLR variable probability distributions for the two hypotheses are Gaussian distributions with means separated by their variance. A good statistical test (such as a Neymann-Pearson criterion with false alarm error of 10^{-4} and detection probability of 0.99) can be achieved for a sufficiently large Λ, typically about 36 or 15.6 dB.

Figure 6.4 illustrates the LLR probability distributions for a small value of Λ. Included in the figure is a threshold defined by the false alarm criterion. Typically, the test is defined such that a given probability is accepted that H0

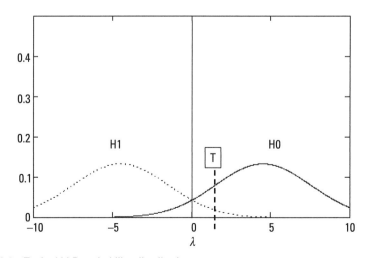

Figure 6.4 Typical LLR probability distributions.

will be estimated as true when in fact hypothesis H1 is true. The area of the H1 distribution to the right of the threshold represents this false alarm probability. The area of the H0 distribution to the right of the threshold represents the probability of detection or acceptance of H0 when indeed hypothesis H0 is true.

From (6.33), the LLR test performance depends on the size of the square of the measurement means difference for the two hypotheses relative to the measurement spread squared or the variance of the measurement probability distribution. The Λ is basically a measure of the signal-to-interference ratio.

For the monopulse ratio measurement, the variance of the noise (σ^2 in [6.33] is assuming angles in radians, an antenna beamwidth [BW], and the JNR).

$$\text{var} = \langle \text{nse}^* \cdot \text{nse} \rangle = \frac{\text{BW}^2}{1.885^2 \cdot \text{JNR}} \quad (6.34)$$

Thus, combining the results of (6.22), (6.33), and (6.34), the test metric is approximately (using the approximation for z in [6.15])

$$|\Lambda| = \frac{\left(\left|\delta\psi \cdot \frac{z}{1+z}\right|\right)^2}{\text{var}} \approx \left|(\psi_S - \psi_J)\right|^2 \cdot \frac{1.885^2}{\text{BW}^2} \cdot \text{SNR} \quad (6.35)$$

The test performance relates to and improves with the angular separation squared between the HVU target and the EA source and with the SNR. This test improves with increasing angular separation between the EA platform and the ship target as long as the target is within the beam. Therefore, the test improves as the range decreases or as the engagement proceeds. The test also improves with increased coherent processing gain. Both increased processing gain and decreased range improve the receiver processed SNR. And the decreasing range also increases the angular separation for a total rate of improvement at about range to the inverse sixth power.

The level of jamming does not degrade the test performance as long as the ASM sensor data dynamic range is adequate to not saturate the ADC or render the target smaller than the ADC least significant bit or two. It is very important to note that as the jamming level increases the difference in the means between H0 and H1 in (6.24) and (6.25) decreases. However, as the jamming level increases the accuracy of the monopulse measurement improves in the same amount (see [6.34]). This results in (6.35), the test performance being generally independent of the jamming level.

These theoretical results were calculated for various SNRs, a very high level of JSR, and a BW of about 10°. The theoretical performance is plotted in Figure 6.5.

Two points are worth repeating. First, the results of this test are not sensitive to the jamming power level. The SNR in (6.35) and Figure 6.5 is the ratio of the HVU echo signal power to the ASM radar receiver noise power in the absence of jamming. For later use in Section 6.3, it is noted that the SNR used in this section relates to the SNR as measured in the Σ channel. In the notation of Chapter 3 and assuming Doppler processing of P pulses prior to monopulse estimation

$$\text{SNR} = \frac{PA_S^2}{\sigma_N^2} \qquad (6.36)$$

Second, this is a very sensitive test to angular separation of the HVU target and the EA source. For example, 0.5° corresponds to about 170m cross-range difference at about 20 km, and 1° corresponds to about 340m at 20 km and about 170m cross-range difference at 10 km. As shown, the test metric is basically a signal measurement level (difference in monopulse ratio means) relative to the interference (monopulse ratio measurement noise variance) level (SIR). This test metric is basically an estimate of the parameter Λ in the LLR formalism. From Figure 6.5, the SIR is about 10 dB for an SNR of 24

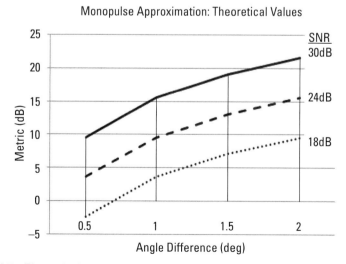

Figure 6.5 Theoretical results for monopulse ratio test.

dB and an angular separation of 1.0°. The SIR is 10 dB for an SNR of 30 dB and an angular separation of about 0.5°. The test metric is about 10 dB for a 15-dB SNR and 2.0° angular separation.

The monopulse ratio has been extensively examined with actual pulsed Doppler radar data. This simple test has been shown to render cover noise EA very transparent. While the cover jamming level is large compared to the signal level in the Σ channel when the two dimensional (Σ and Δ) array of data is converted to monopulse ratio, the Doppler range cell containing the ship target can be readily distinguished from other cells. Again the reason this EP algorithm is successful is because the monopulse ratio is very consistent in all cells that do not contain a target. The stronger the jamming the more consistent the ratio array of data becomes.

Figures 6.6 and 6.7 qualitatively illustrate the concept results. The left image is a standard view of Σ channel data array (amplitude data versus Doppler and range). The array in both cases is heavily corrupted by cover EA and no targets are visible. It is seen that the EA has deliberately varied the noise jamming to create a noise-like-view. The right image in each figure is the same Doppler range array of cells, but using the magnitude of the monopulse ratio at each cell. The circle indicates the location of the ship target (not visible in the left image) that was detected in the monopulse array. Once the HVU is isolated to a particular Doppler range cell the required HVU information can be readily measured.

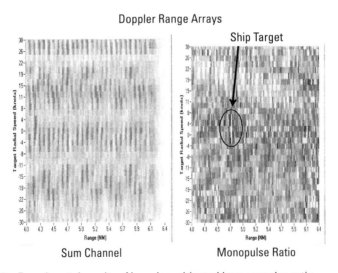

Figure 6.6 Experimental results of jamming mitigated by monopulse ratio.

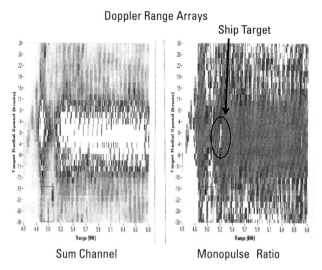

Figure 6.7 Second example of experimental results of jamming mitigated by monopulse ratio.

When the ASM seeker sensor is attacked with noise jamming EA the monopulse ratio data can be used by the ASM processor as an extremely effective EP technique to mitigate the effects of this EA. This array formed from the coherent combination of the data in the two independent receiver data arrays reveals the location of any and all targets that are at an angular direction, different from the source of the EA. This test continuously and rapidly improves with decreasing range.

If a target is in the direction of the EA source the target will eventually burn through the jamming with decreasing range. In addition, other properties of the data based on knowledge of the ASM radar antenna polarization patterns can be used to identify the cell containing the ship target. These algorithms were referenced in Chapter 4.

6.2 ASM STAP Processing

In the previous section, it was shown that detection of the HVU in the presence of cover jamming can be improved by using the two coherent receiver channels rather than depending on only the Σ channel. In that case the two arrays were used to compute the complex monopulse ratio for the Doppler range cells. This can be done for all the Doppler values and for the entire range swath when searching for any targets. If the target location in the array is previously known the ratio needs only to be computed in the region around the target for modified CFAR and target tracking.

Using the full array of data when there is no EA should already improve the detection performance by 3 dB, by simply doubling the amount of data available. It was assumed in the classical processing scheme that the range samples of the Σ channel are subjected to Doppler processing for each CPI. Standard detection techniques such as CFAR processing are used to adaptively detect the target amplitude at each cell in the Doppler range array.

Instead of Doppler processing the CPI of data at each range cell and then conducting a detection algorithm, examine an alternative but equivalent detection scheme. Instead of the P Doppler measurements at each range cell form the array of P time samples (one time sample for each pulse) at each range cell. Use the steering vector (W) as a matched filter for the data sample in the CUT. The detection test can be written as (where the vectors have P elements and the matrices are $P \times P$)

$$\text{H0: } \Sigma(\text{H0}) = A_S \Sigma_S + \sigma_N \Sigma_N \tag{6.37}$$

$$\text{H1: } \Sigma(\text{H1}) = \sigma_N \Sigma_N \tag{6.38}$$

The scalar measurements (Z) are formed by applying the steering vector (W) to the Σ vectors.

$$Z = W^+ \Sigma(Hi) \tag{6.39}$$

If H1 is true a noise measurement results. If H0 is true a signal measurement plus noise results. The optimal test maximizes the magnitude squared of the signal part divided by the noise or interference part.

$$\text{SIR} = \frac{A_S^2 \cdot \Sigma_S^+ W W^+ \Sigma_S}{\sigma_N^2 \cdot \left\langle \Sigma_N^+ W W^+ \Sigma_N \right\rangle} \tag{6.40}$$

Assuming white noise and using the Schwartz Inequality, it is shown that the optimal steering vector is

$$W = k M^{-1} \Sigma_S \tag{6.41}$$

$$M = \left\langle (\sigma_N \Sigma_N)(\sigma_N \Sigma_N)^+ \right\rangle = \sigma_N^2 I \tag{6.42}$$

Multiple Receiver EP Signal Processing

The P components of the Σ_S vector are taken from the dominant terms of (3.81) (where Ω represents the 2×2 radar cross section for the target)

$$\Sigma_S(p) = e^{\left[2\pi i \left\{(\beta\Phi + f_D T)p - \frac{2R_S}{\lambda}\right\}\right]} \cdot \cos^2(2\pi\Phi_S) \cdot \langle \Sigma|\Omega|\Sigma \rangle \qquad (6.43)$$

The matrix M is diagonal for white noise and SIR is the customary SNR. Inserting the results of (6.41) and (6.42) into (6.40) is essentially to perform a Doppler processing that includes the angle term as part of the steering vector. (It is trivial to include filter weighting into this formalism.) The result is

$$\sigma_N^2 \cdot \langle \Sigma_N^+ W W^+ \Sigma_N \rangle = P \cdot \sigma_N^2 \cdot \left[\cos^2(2\pi\Phi) \cdot \langle \Sigma|\Omega|\Sigma \rangle\right]^2 \qquad (6.44)$$

$$A_S^2 \cdot \Sigma_S^+ W W^+ \Sigma_S = P^2 \cdot A_S^2 \cdot \left[\cos^2(2\pi\Phi) \cdot \langle \Sigma|\Omega|\Sigma \rangle\right]^2 \qquad (6.45)$$

$$\text{SIR} = \text{SNR} = \frac{P \cdot A_J^2}{\sigma_N^2} \qquad (6.46)$$

In this standard formalism the exact same processing is executed in both the Σ channel and the Δ channel. All of the detection and other tasks are performed using only the Σ channel data array. The selected data cell of interest in the Σ channel is then combined with the same cell in the Δ channel to form the monopulse ratio. This measurement is then used in the guidance subsystem.

Now consider the time (pulse number), range, and angle array of data for the full array of Σ channel and Δ channel prior to Doppler processing. For every CPI and for each range cell, there are $P\Sigma$ channel complex data values and $P\Delta$ channel data values. For each range cell and prior to Doppler processing, generate the $2P$ component X data vector by alternating X_+ and X_- components where for each pulse p form the complex vector components

$$X_+(p) = \Sigma(p) + \Delta(p) \qquad (6.47)$$

$$X_-(p) = \Sigma(p) - \Delta(p) \qquad (6.48)$$

If the full vector of data is utilized (X vectors; the vectors have $2P$ elements and the matrices are $2P \times 2P$) the detection hypothesis test becomes

$$\text{H0: } X(\text{H0}) = A_S X_S + \sigma_N X_N \qquad (6.49)$$

$$\text{H1: } X(H1) = \sigma_N X_N \qquad (6.50)$$

Again the scalar measurement Z is formed in the same manner as above

$$Z = W^+ X(Hi) \qquad (6.51)$$

Repeating the analysis above: If H1 is true a noise measurement results. If H0 is true a signal measurement plus noise results. The optimal test maximizes the magnitude squared of the signal part divided by the noise interference part.

$$\text{SIR} = \frac{A_S^2 \cdot X_S^+ W W^+ X_S}{\sigma_N^2 \cdot \langle X_N^+ W W^+ X_N \rangle} \qquad (6.52)$$

Assuming white noise and using the Schwartz Inequality, it is shown that the optimal steering vector is

$$W = k M^{-1} X_S \qquad (6.53)$$

$$M = \langle (\sigma_N X_N)(\sigma_N X_N)^+ \rangle = \sigma_N^2 I \qquad (6.54)$$

In (6.40), an approximation for the Σ channel components was used from (3.81). Assume the same approximation from (3.82). The corresponding Δ components are

$$\Delta_S(p) = e^{\left[2\pi i\left\{(\beta\Phi + f_D T)p - \frac{2R_0}{\lambda}\right\}\right]} \cdot i\cos(2\pi\Phi)\sin(2\pi\Phi) \cdot \langle \Sigma|\Omega|\Sigma \rangle \qquad (6.55)$$

Inserting (6.43) and (6.55) into the definitions for (6.47) and (6.48) gives the $2P$ components of the X vector

$$X_{S+}(p) = e^{\left[2\pi i\left\{(\beta\Phi + f_D T)p - \frac{2R_0}{\lambda}\right\}\right]} \cdot e^{+2\pi i\Phi} \cdot \cos(2\pi\Phi) \cdot \langle \Sigma|\Omega|\Sigma \rangle \qquad (6.56)$$

$$X_{S-}(p) = e^{\left[2\pi i\left\{(\beta\Phi + f_D T)p - \frac{2R_0}{\lambda}\right\}\right]} \cdot e^{-2\pi i\Phi} \cdot \cos(2\pi\Phi) \cdot \langle \Sigma|\Omega|\Sigma \rangle \qquad (6.57)$$

Using the model for the X vector components the terms in (6.52) can be evaluated as before

$$\sigma_N^2 \cdot \langle X_N^+ W W^+ X_N \rangle = 2P \cdot \sigma_N^2 \cdot \left[\cos(2\pi\Phi) \cdot \langle \Sigma|\Omega|\Sigma\rangle\right]^2 \quad (6.58)$$

$$A_S^2 \cdot X_S^+ W W^+ X_S = 4P^2 \cdot A_S^2 \cdot \left[\cos(2\pi\Phi) \cdot \langle \Sigma|\Omega|\Sigma\rangle\right]^2 \quad (6.59)$$

Again the expression for the SIR in (6.52) becomes the SNR.

$$\text{SIR} = \text{SNR} = \frac{2P \cdot A_J^2}{\sigma_N^2} \quad (6.60)$$

Comparing this result with the expression in (6.46), it is seen that this SNR is 3 dB greater than above as a consequence of using twice as much data in the detection process. Also, it is noted that the steering vector is not simply the Doppler filter transform. The steering vector in (6.53) also includes the antenna angle to the target through the second exponential in (6.56) and (6.57). The importance of this term will become clearer in the following section.

For now, consider again the case of multiple false targets. Repeating the full expressions for the hypothesis test at the CUT

$$\text{H0: } X(\text{H0}) = A_S X_S + \sigma_N X_N \quad (6.61)$$

$$\text{H1: } X(\text{H1}) = A_J X_J + \sigma_N X_N \quad (6.62)$$

The X's on the right-hand side are given in Chapter 3 and are repeated here for completeness. The $2P$ noise X samples are assumed to be (where η is a random phase and k is the component number from 1 to $2P$)

$$X_N(k) = \exp\left(2\pi i \eta_k^N\right) \quad (6.63)$$

The signal components are P pairs of samples where the variables refer to ship geometry

$$X_{S+}(p) = \exp\left[2\pi i(\beta\Phi + f_D T)p\right] \cdot \exp(2\pi i\Phi) \cdot S_+ \quad (6.64)$$

$$X_{S-}(p) = \exp\left[2\pi i(\beta\Phi + f_D T)p\right] \cdot \exp(-2\pi i\Phi) \cdot S_- \quad (6.65)$$

The coefficients (S) are given as either of the expressions

$$S_+ = \cos 2\pi\Phi\big[\langle\Sigma|\Omega|\Sigma\rangle + \langle\Delta|\Omega|\Sigma\rangle\big]$$
$$+ i\sin 2\pi\Phi\big[\langle\Sigma|\Omega|\Delta\rangle + \langle\Delta|\Omega|\Delta\rangle\big] \qquad (6.66)$$

$$S_- = \cos 2\pi\Phi\big[\langle\Sigma|\Omega|\Sigma\rangle - \langle\Delta|\Omega|\Sigma\rangle\big]$$
$$+ i\sin 2\pi\Phi\big[\langle\Sigma|\Omega|\Delta\rangle - \langle\Delta|\Omega|\Delta\rangle\big] \qquad (6.67)$$

$$2S_+ = \exp(2\pi i\Phi)\cdot\big[\langle\Sigma|\Omega|\Sigma\rangle + \langle\Delta|\Omega|\Sigma\rangle + \langle\Sigma|\Omega|\Delta\rangle + \langle\Delta|\Omega|\Delta\rangle\big]$$
$$+ \exp(-2\pi i\Phi)\big[\langle\Sigma|\Omega|\Sigma\rangle + \langle\Delta|\Omega|\Sigma\rangle - \langle\Sigma|\Omega|\Delta\rangle - \langle\Delta|\Omega|\Delta\rangle\big] \qquad (6.68)$$

$$2S_- = \exp(2\pi i\Phi)\cdot\big[\langle\Sigma|\Omega|\Sigma\rangle - \langle\Delta|\Omega|\Sigma\rangle + \langle\Sigma|\Omega|\Delta\rangle - \langle\Delta|\Omega|\Delta\rangle\big]$$
$$+ \exp(-2\pi i\Phi)\big[\langle\Sigma|\Omega|\Sigma\rangle - \langle\Delta|\Omega|\Sigma\rangle - \langle\Sigma|\Omega|\Delta\rangle + \langle\Delta|\Omega|\Delta\rangle\big]\} \qquad (6.69)$$

It is noted that the basic form of the expressions are identical to those of side-looking SAR data processed via STAP techniques. These techniques have been explored by PRC personnel for the case of two receiver channels and near-forward looking radar [1–5]. While the normal STAP results degrade when reduced to two channels and when looking near forward, the results are enhanced when the platform speed is increased. It was noted at the Naval Research Laboratory that these are exactly the conditions applicable to the modern ASM scenario [6].

The jamming samples are again P pairs of samples in which the geometry variables refer to the jammer platform geometry

$$X_{J+}(p) = \exp\big[2\pi i(\beta\Phi + fT)p\big]\cdot \exp(2\pi i\Phi)\cdot J_+ \qquad (6.70)$$

$$X_{J-}(p) = \exp\big[2\pi i(\beta\Phi + fT)p\big]\cdot \exp(-2\pi i\Phi)\cdot J_- \qquad (6.71)$$

The coefficients (J) are given as either of the expressions

$$J_+ = \cos 2\pi\Phi\big[\langle\Sigma|J\rangle\langle J|\Sigma\rangle + \langle\Delta|J\rangle\langle J|\Sigma\rangle\big]$$
$$+ i\sin 2\pi\Phi\big[\langle\Sigma|J\rangle\langle J|\Delta\rangle + \langle\Delta|J\rangle\langle J|\Delta\rangle\big] \qquad (6.72)$$

$$J_- = \cos 2\pi\Phi\big[\langle\Sigma|J\rangle\langle J|\Sigma\rangle - \langle\Delta|J\rangle\langle J|\Sigma\rangle\big]$$
$$+ i\sin 2\pi\Phi\big[\langle\Sigma|J\rangle\langle J|\Delta\rangle - \langle\Delta|J\rangle\langle J|\Delta\rangle\big] \qquad (6.73)$$

$$2J_+ = \exp(2\pi i\Phi)\begin{bmatrix}\langle\Sigma|J\rangle\langle J|\Sigma\rangle + \langle\Delta|J\rangle\langle J|\Sigma\rangle \\ +\langle\Sigma|J\rangle\langle J|\Delta\rangle + \langle\Delta|J\rangle\langle J|\Delta\rangle\end{bmatrix}$$
$$+\exp(-2\pi i\Phi)\begin{bmatrix}\langle\Sigma|J\rangle\langle J|\Sigma\rangle + \langle\Delta|J\rangle\langle J|\Sigma\rangle \\ -\langle\Sigma|J\rangle\langle J|\Delta\rangle - \langle\Delta|J\rangle\langle J|\Delta\rangle\end{bmatrix} \quad (6.74)$$

$$2J_- = \exp(2\pi i\Phi)\begin{bmatrix}\langle\Sigma|J\rangle\langle J|\Sigma\rangle - \langle\Delta|J\rangle\langle J|\Sigma\rangle \\ +\langle\Sigma|J\rangle\langle J|\Delta\rangle - \langle\Delta|J\rangle\langle J|\Delta\rangle\end{bmatrix}$$
$$+\exp(-2\pi i\Phi)\begin{bmatrix}\langle\Sigma|J\rangle\langle J|\Sigma\rangle - \langle\Delta|J\rangle\langle J|\Sigma\rangle \\ -\langle\Sigma|J\rangle\langle J|\Delta\rangle + \langle\Delta|J\rangle\langle J|\Delta\rangle\end{bmatrix} \quad (6.75)$$

Correlation Function Approach

One approach proposed in the PRC references and discussed above is to examine the correlation between pairs of targets. Consider the correlation between a pair of two distinct false targets from the same EA system.

$$\langle X_{H1}^+ X_{H1}'\rangle = PA_J^2\left(|J_+^2| + |J_-^2|\right) \quad (6.76)$$

Assuming the dominant terms in (6.65) and (6.66) are the first terms, this is approximately

$$\langle X_{H1}^+ X_{H1}'\rangle = 2PA_J^2 K_J \cos^2(2\pi\Phi_J) \quad (6.77)$$

Normalizing this expression and realizing that all of the false targets are from the same direction (angle Φ), the correlation is equal to 1.

$$\frac{\langle X_{H1}^+ X_{H1}'\rangle}{\sqrt{\langle X_{H1}'^+ X_{H1}'\rangle \cdot \langle X_{H1}^+ X_{H1}\rangle}} = 1 \quad (6.78)$$

If one of the targets is the ship (with no false target in the CUT)

$$\langle X_{H0}^+ X_{H1}'\rangle = PA_J A_S \operatorname{sinc}\left(\pi\beta[\Phi_S - \Phi_J]P\right)$$
$$\left(J_+^* S_+ \exp\{2\pi i[\Phi_S - \Phi_J]\}\right. \quad (6.79)$$
$$\left. + J_-^* S_- \exp\{-2\pi i[\Phi_S - \Phi_J]\}\right)$$

Using the dominant terms for the J terms and the S terms gives approximately

$$\langle X_{H0}^+ X_{H1}' \rangle = 2PA_J A_S \operatorname{sinc}\left(\pi\beta\left[\Phi_S - \Phi_J\right]P\right) \\ \sqrt{K_J K_S} \cdot \cos\left[2\pi\left(\Phi_S - \Phi_J\right)\right]) \tag{6.80}$$

Assuming the false targets are approximately the same magnitude as real targets and normalizing this expression gives a value of magnitude less than 1 when the EA system is not on the target platform. The complete expression for the correlation between a true target and a false target is

Correlation =

$$\frac{\operatorname{sinc}\left(\pi\beta\left[\Phi_S - \Phi_J\right]P\right)\left(J_+^* S_+ \exp\{2\pi i\left[\Phi_S - \Phi_J\right]\} + J_-^* S_- \exp\{-2\pi i\left[\Phi_S - \Phi_J\right]\}\right)}{\sqrt{(|J_+^2| + |J_-^2|)(|S_+^2| + |S_-^2|)}}$$

(6.81)

Using the dominant terms again and assuming the angles are small, it is seen that

$$\text{Correlation} \approx \operatorname{sinc}\left(\pi\beta\left[\Phi_S - \Phi_J\right]P\right) \tag{6.82}$$

From Chapter 3, it is recalled that

$$\Phi = \frac{d}{2\lambda} \cdot \sin(\psi) \tag{6.83}$$

$$\beta = \frac{-4vT\sin\gamma}{\lambda} \tag{6.84}$$

Thus, the correlation between a false target and the true target is approximately

$$\text{Correlation} \approx \operatorname{sinc}\left(2\pi\frac{vTd}{\lambda^2}\left[\psi_S - \psi_J\right]P\right) \tag{6.85}$$

An alternative approach was discussed above. In [2, 3], the authors solve for the polarization Jones vector given the ASM antenna patterns. (Note: if there is a false target in the cell with the target, the authors show that the superimposed expression is more complicated, but tractable.) This algorithm approach is conjectured by the authors to succeed even when the EA is at the ship angle. The references compare several similar algorithms.

Additionally, it is noted that the *X* data vectors can be more easily collected from the simpler system shown in Figure 6.8 compared to Figure 6.1. In this new configuration, there is no need for the antenna hybrid since the U and L antenna outputs correspond to the components previously formed from adding and subtracting the Σ and Δ channel results. Also, it is noted that the data in the two receivers are more similar in level and other characteristics than in the previous sensor configuration. All of this should make for a more efficient and easier to implement hardware seeker system.

6.3 Cover Jamming EP

Now consider the more difficult case of cover or noise jamming EA. Again the *X* data vectors can be more easily collected from the simpler system shown in Figure 6.8. The full expressions for the hypothesis test at the range CUT are (for simplicity suppress the subscript *N* from the noise level σ_N for the remainder of this chapter)

$$\text{H0: } X(\text{H0}) = A_S X_S + A_J X_J + \sigma X_N \qquad (6.86)$$

$$\text{H1: } X(\text{H1}) = A_J X_J + \sigma X_N \qquad (6.87)$$

These are the data vectors for the two receiver channels for a single range cell after *P* transmit pulses are combined into a single data sample vector. The *X*'s on the right-hand side are given above and are repeated here for completeness. The 2*P* noise *X* samples are (where again *k* refers to the component number: *k* = 1 to 2*P*)

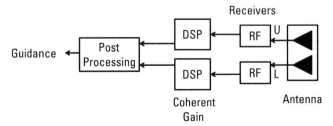

Figure 6.8 Modern ASM sensor configuration.

$$X_N(k) = \exp(2\pi i \eta_k^N) \tag{6.88}$$

The signal samples are P pairs of samples in which the variables f_D and Φ refer to ship geometry (the p refers to pulse number and vector component pairs)

$$X_{S+}(p) = \exp\left[2\pi i(\beta\Phi + f_D T)p\right] \cdot \exp(2\pi i\Phi) \cdot S_+ \tag{6.89}$$

$$X_{S-}(p) = \exp\left[2\pi i(\beta\Phi + f_D T)p\right] \cdot \exp(-2\pi i\Phi) \cdot S_- \tag{6.90}$$

As before the S coefficients are given as either of the expressions

$$\begin{aligned}S_+ &= \cos 2\pi\Phi\left[\langle\Sigma|\Omega|\Sigma\rangle + \langle\Delta|\Omega|\Sigma\rangle\right] \\ &+ i\sin 2\pi\Phi\left[\langle\Sigma|\Omega|\Delta\rangle + \langle\Delta|\Omega|\Delta\rangle\right]\end{aligned} \tag{6.91}$$

$$\begin{aligned}S_- &= \cos 2\pi\Phi\left[\langle\Sigma|\Omega|\Sigma\rangle - \langle\Delta|\Omega|\Sigma\rangle\right] \\ &+ i\sin 2\pi\Phi\left[\langle\Sigma|\Omega|\Delta\rangle - \langle\Delta|\Omega|\Delta\rangle\right]\end{aligned} \tag{6.92}$$

$$\begin{aligned}2S_+ &= \exp(2\pi i\Phi) \cdot \left[\langle\Sigma|\Omega|\Sigma\rangle + \langle\Delta|\Omega|\Sigma\rangle + \langle\Sigma|\Omega|\Delta\rangle + \langle\Delta|\Omega|\Delta\rangle\right] \\ &+ \exp(-2\pi i\Phi)\left[\langle\Sigma|\Omega|\Sigma\rangle + \langle\Delta|\Omega|\Sigma\rangle - \langle\Sigma|\Omega|\Delta\rangle - \langle\Delta|\Omega|\Delta\rangle\right]\end{aligned} \tag{6.93}$$

$$\begin{aligned}2S_- &= \exp(2\pi i\Phi) \cdot \left[\langle\Sigma|\Omega|\Sigma\rangle - \langle\Delta|\Omega|\Sigma\rangle + \langle\Sigma|\Omega|\Delta\rangle - \langle\Delta|\Omega|\Delta\rangle\right] \\ &+ \exp(-2\pi i\Phi)\left[\langle\Sigma|\Omega|\Sigma\rangle - \langle\Delta|\Omega|\Sigma\rangle - \langle\Sigma|\Omega|\Delta\rangle + \langle\Delta|\Omega|\Delta\rangle\right]\end{aligned} \tag{6.94}$$

The jamming samples are also P pairs of samples in which the variables Φ refer to the jammer platform geometry

$$X_{J+}(p) = \exp\left[2\pi i\left(\beta\Phi p + i\eta_p^J\right)\right] \cdot \exp(2\pi i\Phi) \cdot J_+ \tag{6.95}$$

$$X_{J-}(p) = \exp\left[2\pi i\left(\beta\Phi p + i\eta_p^J\right)\right] \cdot \exp(-2\pi i\Phi) \cdot J_- \tag{6.96}$$

The coefficients J are given as either of the expressions

$$\begin{aligned}J_+ &= \cos 2\pi\Phi\left[\langle\Sigma|J\rangle\langle J|\Sigma\rangle + \langle\Delta|J\rangle\langle J|\Sigma\rangle\right] \\ &+ i\sin 2\pi\Phi\left[\langle\Sigma|J\rangle\langle J|\Delta\rangle + \langle\Delta|J\rangle\langle J|\Delta\rangle\right]\end{aligned} \tag{6.97}$$

$$J_- = \cos 2\pi\Phi\big[\langle\Sigma|J\rangle\langle J|\Sigma\rangle - \langle\Delta|J\rangle\langle J|\Sigma\rangle\big]$$
$$+ i\sin 2\pi\Phi\big[\langle\Sigma|J\rangle\langle J|\Delta\rangle - \langle\Delta|J\rangle\langle J|\Delta\rangle\big] \quad (6.98)$$

$$2J_+ = \exp(2\pi i\Phi)\begin{bmatrix}\langle\Sigma|J\rangle\langle J|\Sigma\rangle + \langle\Delta|J\rangle\langle J|\Sigma\rangle \\ + \langle\Sigma|J\rangle\langle J|\Delta\rangle + \langle\Delta|J\rangle\langle J|\Delta\rangle\end{bmatrix}$$
$$+ \exp(-2\pi i\Phi)\begin{bmatrix}\langle\Sigma|J\rangle\langle J|\Sigma\rangle + \langle\Delta|J\rangle\langle J|\Sigma\rangle \\ - \langle\Sigma|J\rangle\langle J|\Delta\rangle - \langle\Delta|J\rangle\langle J|\Delta\rangle\end{bmatrix} \quad (6.99)$$

$$2J_- = \exp(2\pi i\Phi)\begin{bmatrix}\langle\Sigma|J\rangle\langle J|\Sigma\rangle - \langle\Delta|J\rangle\langle J|\Sigma\rangle \\ + \langle\Sigma|J\rangle\langle J|\Delta\rangle - \langle\Delta|J\rangle\langle J|\Delta\rangle\end{bmatrix}$$
$$+ \exp(-2\pi i\Phi)\begin{bmatrix}\langle\Sigma|J\rangle\langle J|\Sigma\rangle - \langle\Delta|J\rangle\langle J|\Sigma\rangle \\ - \langle\Sigma|J\rangle\langle J|\Delta\rangle + \langle\Delta|J\rangle\langle J|\Delta\rangle\end{bmatrix} \quad (6.100)$$

In classic pulsed Doppler processing, summarized above, the steering vectors are simply the various Doppler filters. This result was found in the previous section to correspond to the assumption of detection of a ship signal or false target signal in white noise. That is, the noise interference is white and Gaussian when no cover EA is present.

For the case herein, the hypotheses of (6.86) and (6.87) both contain a nonwhite interference term. The preferred method is to process these signals via a prewhitened steering vector W as in standard detection theory and standard STAP processing [5, 6]. Define the measurement Z for the hypothesis test defined in (6.86) and (6.87)

$$Z = W^+ X_H \quad (6.101)$$

If H1 is true, a noise plus EA interference measurement results. If H0 is true, a signal measurement plus noise and interference results. The optimal test maximizes the magnitude squared of the signal part divided by the noise plus interference part.

$$\text{SIR} = \frac{A_S^2 \cdot X_S^+ W W^+ X_S}{\langle X_{H1}^+ W W^+ X_{H1}\rangle} \quad (6.102)$$

Assume a result similar to the results above. The definition of X in this expression will be defined below. The optimal steering vector is of the form

$$W = kM^{-1}X \qquad (6.103)$$

As above, matrix M is the expectation of the interference plus noise squared or the denominator in (6.102)

$$M = \langle (A_J X_J + \sigma X_N)(A_J X_J + \sigma X_N)^{+} \rangle \qquad (6.104)$$

Using the expressions for X_J and X_N the $2P \times 2P$ matrix M can be computed. This matrix is all zeroes except for the 2×2 matrices H along the diagonal. The matrices H are

$$H = \begin{bmatrix} (A_j^2|J_+^2| + \sigma^2) & A_j^2 \exp(2\pi i \cdot 2\Phi_J) J_+ J_-^* \\ A_j^2 \exp(-2\pi i \cdot 2\Phi_J) J_+^* J_- & (A_j^2|J_-^2| + \sigma^2) \end{bmatrix} \qquad (6.105)$$

Likewise, the inverse of M is all zeroes except for the 2×2 matrices along the diagonal. These submatrices are the inverse of H

$$H^{-1} = \begin{bmatrix} (A_j^2|J_-^2| + \sigma^2) & -A_j^2 \exp(2\pi i \cdot 2\Phi_J) J_+ J_-^* \\ -A_j^2 \exp(-2\pi i \cdot 2\Phi_J) J_+^* J_- & (A_j^2|J_+^2| + \sigma^2) \end{bmatrix}$$
$$\cdot \left[\sigma^2 \cdot \left(A_j^2 \{|J_+^2| + |J_-^2|\} + \sigma^2 \right) \right]^{-1} \qquad (6.106)$$

Examination of Terms with Jamming Steering Filter

Return now to (6.103) and seek a steering vector W that selects the jamming platform. That is, use X_J for the X vector in (6.103). At this point the calculations can be shortened by exploiting the definition of M. Using the jamming steering filter and (6.104) and (6.106), assume $W_+ = J_+$ and $W_- = J_-$ and leave the angle arbitrary for the moment

$$W = kM^{-1}X \qquad (6.107)$$

$$X_+(p) = \exp\left[2\pi i(\beta \Phi_W p)\right] \cdot \exp(2\pi i \Phi_W) \cdot J_+ \qquad (6.108)$$

$$X_-(p) = \exp\left[2\pi i(\beta \Phi_W p)\right] \cdot \exp(-2\pi i \Phi_W) \cdot J_- \qquad (6.109)$$

Processing the interference term (or assuming H1 is true) gives a measurement with the following properties:

$$Z_I = k\left[X^+ M^{-1}\left(A_J X_J + \sigma X_N\right)\right] \quad (6.110)$$

$$\langle Z_I \rangle = 0 \quad (6.111)$$

$$\langle Z_I^* Z_I \rangle = \frac{k^2 P\left[4A_J^2 |J_+^2||J_-^2|\sin^2\left[2\pi\left(\Phi_J - \Phi_W\right)\right] + \sigma^2\{|J_+^2| + |J_-^2|\}\right]}{\sigma^2\left(A_J^2\{|J_+^2| + |J_-^2|\} + \sigma^2\right)} \quad (6.112)$$

Note the jamming is mitigated at the jamming angle (when $\Phi_W = \Phi_J$) because of the definition of M as expected from STAP results. Now examine the ship return part of the vector when H0 is true

$$Z_S = \frac{k A_S P[Q]}{\sigma^2\left(A_J^2\{|J_+^2| + |J_-^2|\} + \sigma^2\right)} \quad (6.113)$$

$$\begin{aligned}[Q] = {} & \exp 2\pi i\left(\Phi_S - \Phi_W\right)\left(A_J^2|J_-^2| + \sigma^2\right) S_+ J_+^* \\
& + \exp\left[-2\pi i\left(\Phi_S - \Phi_W\right)\right]\left(A_J^2|J_+^2| + \sigma^2\right) S_- J_-^* \\
& - A_J^2[|J_-^2| S_+ J_+^* \exp 2\pi i\left(\Phi_S + \Phi_W - 2\Phi_J\right) \\
& + |J_+^2| S_- J_-^* \exp\left[-2\pi i\left(\Phi_S + \Phi_W - 2\Phi_J\right)\right]\end{aligned} \quad (6.114)$$

Now consider two special cases for these expressions.

Case $\Phi_W = \Phi_J$

The cover jamming is canceled in the same manner as for adaptive null steering antenna processing with

$$\langle Z_I^* Z_I \rangle = \frac{k^2 P\{|J_+^2| + |J_-^2|\}}{\left(A_J^2\{|J_+^2| + |J_-^2|\} + \sigma^2\right)} \quad (6.115)$$

$$Z_S = \frac{k A_S P[Q]}{\left(A_J^2\{|J_+^2| + |J_-^2|\} + \sigma^2\right)} \quad (6.116)$$

$$[Q] = \exp 2\pi i \left(\Phi_S - \Phi_J\right) S_+ J_+^* + \exp\left[-2\pi i \left(\Phi_S - \Phi_J\right) S_- J_-^*\right] \quad (6.117)$$

Thus, both of these terms are very small when steering at the jamming platform angle.

Case $\Phi_W = \Phi_S$

Now consider the steering vector adapted to the jamming signal, but pointed at the target ship. In this case

$$\langle Z_I^* Z_I \rangle = \frac{k^2 P\left[4A_J^2 |J_+^2||J_-^2| \sin^2 2\pi\left(\Phi_J - \Phi_S\right) + \sigma^2\left\{|J_+^2| + |J_-^2|\right\}\right]}{\sigma^2\left(A_J^2\left\{|J_+^2| + |J_-^2|\right\} + \sigma^2\right)} \quad (6.118)$$

$$Z_S = \frac{kA_S P[Q]}{\sigma^2\left(A_J^2\left\{|J_+^2| + |J_-^2|\right\} + \sigma^2\right)} \quad (6.119)$$

$$[Q] = \sigma^2\left(S_+ J_+^* + S_- J_-^*\right)$$
$$+ 2A_J^2\left[\left(|J_-^2| S_+ J_+^* + |J_+^2| S_- J_-^*\right) \sin^2 2\pi\left(\Phi_S - \Phi_J\right)\right. \quad (6.120)$$
$$\left. - i\left(|J_-^2| S_+ J_+^* - |J_+^2| S_- J_-^*\right) \sin 2\pi\left(\Phi_S - \Phi_J\right) \cos 2\pi\left(\Phi_S - \Phi_J\right)\right]$$

The expressions for these two cases are similar to the previous results, except that now the EA has been more significantly mitigated.

Case $\Phi_W = \Phi_S$ and Dominant Terms Only

To more clearly see this result, assume the expressions for the antenna terms using only the dominant terms. Most of these terms can be absorbed into the definitions for the amplitude. The expressions are greatly simplified especially when the jamming level is very high. In the presence of high-level cover jamming

$$\langle Z_I^* Z_I \rangle = \frac{2k^2 P \sin^2\left[2\pi\left(\Phi_J - \Phi_S\right)\right]}{\sigma^2} \quad (6.121)$$

$$Z_S = \frac{2kA_S P \sin^2\left[2\pi\left(\Phi_S - \Phi_J\right)\right]}{\sigma^2} \quad (6.122)$$

$$\text{SIR} = \frac{2A_S^2 P \sin^2\left[2\pi(\Phi_S - \Phi_J)\right]}{\sigma^2} \qquad (6.123)$$

$$\text{SIR} \approx \frac{2A_S^2 P \left[2\pi(\Phi_S - \Phi_J)\right]^2}{\sigma^2} = \text{SNR} \cdot \left[2\pi(\Phi_S - \Phi_J)\right]^2 \qquad (6.124)$$

These results are very similar to the results in Section 6.1 in (6.35). The SIR expression depends on the SNR. This SNR is 3 dB higher than the previous result for the Σ channel SNR. The SIR expression depends on the angular separation between the ship target and the EA platform.

Target Steering Filter and Classical Detection

Consider again the classical detection formalism. The full expressions for the hypothesis test at the CUT are

$$\text{H0: } X(H0) = A_S X_S + A_J X_J + \sigma X_N \qquad (6.125)$$

$$\text{H1: } X(H1) = A_J X_J + \sigma X_N \qquad (6.126)$$

Again, shorten the calculations by exploiting the definition of M. Assume the optimal steering vector but this time use the proper steering vector for the optimal target detection: $W_+ = S_+$ and $W_- = S_-$ and use a Doppler value and angle for the steering vector corresponding to the target (f_D and Φ_S).

$$Z = W^+ X_H \qquad (6.127)$$

$$W = kM^{-1} X_S \qquad (6.128)$$

$$X_{S+}(p) = \exp\left[2\pi i(\beta\Phi_S + f_D T)p\right] \cdot \exp(2\pi i \Phi_S) \cdot S_+ \qquad (6.129)$$

$$X_{S-}(p) = \exp\left[2\pi i(\beta\Phi_S + f_D T)p\right] \cdot \exp(-2\pi i \Phi_S) \cdot S_- \qquad (6.130)$$

Define the following terms when hypothesis H1 is true:

$$Z_1 = k\left[X_S^+ M^{-1}(A_J X_J + \sigma X_N)\right] \qquad (6.131)$$

$$\langle Z_1 \rangle = 0 \qquad (6.132)$$

$$\text{Variance}_1 = \langle Z_1^* Z_1 \rangle = k^2 X_S^+ M^{-1} X_S \qquad (6.133)$$

$$\langle Z_1^* Z_1 \rangle = \left\{ \sigma^2 \left(A_J^2 \{|J_+^2| + |J_-^2|\} + \sigma^2 \right) \right\}^{-1}$$
$$\cdot \left\{ k^2 P \left[J^2 \{ (|S_+^2\|J_-^2| + |S_-^2\|J_+^2|) \right. \right.$$
$$- \left(S_+^* S_- J_+ J_-^* \exp 2\pi i 2 (\Phi_J - \Phi_S) \right) \qquad (6.134)$$
$$+ S_-^* S_+ J_- J_+^* \exp \left[-2\pi i 2 (\Phi_J - \Phi_S) \right] \right\}$$
$$+ \sigma^2 \{|J_+^2| + |J_-^2|\} \right\}$$

Notice the expected value of this measurement is 0 in range cells that contain only EA. That is, the expected value of the measurement is 0 in range cells that do not contain the HVU target. The variance of this measurement is given by (6.133) and (6.134).

Processing the interference part of the measurement when H0 is true gives the same values

$$Z_I = k \left[X_S^+ M^{-1} (A_J X_J + \sigma X_N) \right] \qquad (6.135)$$

$$\langle Z_I \rangle = 0 \qquad (6.136)$$

$$\langle Z_I^* Z_I \rangle = k^2 X_S^+ M^{-1} X_S \qquad (6.137)$$

$$\langle Z_I^* Z_I \rangle = \left\{ \sigma^2 \left(A_J^2 \{|J_+^2| + |J_-^2|\} + \sigma^2 \right) \right\}^{-1}$$
$$\cdot \left\{ k^2 P \left[A_J^{\;2} \{ (|S_+^2\|J_-^2| + |S_-^2\|J_+^2|) \right. \right.$$
$$- \left(S_+^* S_- J_+ J_-^* \exp 2\pi i 2 (\Phi_J - \Phi_S) \right) \qquad (6.138)$$
$$- S_-^* S_+ J_- J_+^* \exp \left[-2\pi i 2 (\Phi_J - \Phi_S) \right] \right\}$$
$$+ \sigma^2 \{|J_+^2| + |J_-^2|\} \right\}$$

Now define the parameter Λ. Recall that the variance of LLR is the difference in the means squared divided by the interference variance. Thus

$$\Lambda = \frac{k^2 A_S^2 (X_S^+ M^{-1} X_S)^2}{\langle Z_I^* Z_I \rangle} \tag{6.139}$$

$$\Lambda = \frac{A_S^2}{k^2} \langle Z_I^* Z_I \rangle \tag{6.140}$$

$$\Lambda = \left\{ \sigma^2 \left(A_J^2 \{|J_+^2| + |J_-^2|\} + \sigma^2 \right) \right\}^{-1}$$
$$\cdot \left\{ A_S^2 P [A_J^2 \{(|S_+^2\|J_-^2| + |S_-^2\|J_+^2|) - \right.$$
$$- \left(S_+^* S_- J_+ J_-^* \exp 2\pi i 2 (\Phi_J - \Phi_S) \right) \tag{6.141}$$
$$\left. - S_-^* S_+ J_- J_+^* \exp\left[-2\pi i 2 (\Phi_J - \Phi_S)\right]\right\} + \sigma^2 \{|J_+^2| + |J_-^2|\} \right\}$$

The remaining terms when hypothesis H0 is true are

$$Z_0 = k \left[X_S^+ M^{-1} \left(A_S X_S + A_J X_J + \sigma X_N \right) \right] \tag{6.142}$$

$$\langle Z_0 \rangle = k A_S X_S^+ M^{-1} X_S \tag{6.143}$$

$$\text{Variance}_0 = \langle Z_0^* Z_0 \rangle = k^2 X_S^+ M^{-1} X_S \tag{6.144}$$

This measurement has the same variance as Z_1, but with a nonzero mean. Thus, assuming Gaussian statistics, the optimal detection formalism corresponds again to forming the LLR, where subscript "d" refers to the X data for H0 or H1 being true

$$\lambda_d = A_S X_S^+ M^{-1} X_d - \frac{\Lambda}{2} \tag{6.145}$$

$$\langle \lambda_0 \rangle = \frac{\Lambda}{2} \tag{6.146}$$

$$\langle \lambda_1 \rangle = -\frac{\Lambda}{2} \tag{6.147}$$

$$\text{Variance } \lambda_0 = \text{Variance } \lambda_1 = \Lambda = A_S^2 X_S^+ M^{-1} X_S \qquad (6.148)$$

As before the LLR probability distributions are fully defined by the expression for Λ. Simplifying by assuming the coefficients for A_J and A_S are the dominant terms and in the presence of strong EA (high JNR or JSR) the expression simplifies as before.

$$S_+ = S_- = \cos 2\pi \Phi_S \langle \Sigma | \Omega | \Sigma \rangle \qquad (6.149)$$

$$J_+ = J_- = \cos 2\pi \Phi_J \langle \Sigma | J \rangle \langle J | \Sigma \rangle \qquad (6.150)$$

Therefore, for small angles absorb the radar cross section and antenna gains for the ASM antenna and EA antenna into the A_S and A_J terms for convenience

$$\text{SIR} = \Lambda = \frac{2 A_S^2 P \left[2 A_J^2 \sin^2 2\pi (\Phi_J - \Phi_S) + \sigma^2 \right]}{\sigma^2 (2 A_J^2 + \sigma^2)}$$

$$\approx \frac{2 A_S^2 P \sin^2 2\pi (\Phi_J - \Phi_S)}{\sigma^2} \approx \frac{2 A_S^2 P \left[2\pi (\Phi_J - \Phi_S) \right]^2}{\sigma^2} \qquad (6.151)$$

$$= \text{SNR} \cdot \left[2\pi (\Phi_S - \Phi_J) \right]^2$$

$$\text{SIR} \approx \text{SNR} \cdot \left[2\pi (\Phi_S - \Phi_J) \right]^2 \qquad (6.152)$$

This again is the expression for the SIR resulting from the prewhitened matched filter approach. By using the optimal prewhitened matched filter or the jamming signal filter, an optimal detection scheme identifies the CUT as having either jamming or jamming plus ship target. The scalar LLR measurement is formed and the value compared with statistics based on the LLR variance.

This EP performance depends on the underlying SNR of the entire $2P$ dimensional array (which is 3 dB greater than the Σ channel only SNR) and the angular separation between the ship and EA platform. Also, recall that this SNR contains all of the coherent gain from pulse compression and Doppler processing. The EP performance is relatively insensitive to the JSR as long as the ASM receiver (ADC) dynamic range is adequate to not result in corrupted signals and both targets are within the main antenna beam.

This has been a lot to absorb. The reason for these conclusions is based on the form of the expressions for X being the same as the standard expressions for the data in the STAP formalism. The EA (cover jamming or false targets) and the HVU target echo are expressed in the form of a full $2P$ vector with components

$$X_{S+}(p) = \exp\left[2\pi i(\beta\Phi_S + f_D T)p\right]\cdot\exp(2\pi i\Phi_S)\cdot S_+ \quad (6.153)$$

$$X_{S-}(p) = \exp\left[2\pi i(\beta\Phi_S + f_D T)p\right]\cdot\exp(-2\pi i\Phi_S)\cdot S_- \quad (6.154)$$

$$X_{J+}(p) = \exp\left[2\pi i(\beta\Phi_J p + \eta_p^J)\right]\cdot\exp(2\pi i\Phi_J)\cdot J_+ \quad (6.155)$$

$$X_{J-}(p) = \exp\left[2\pi i(\beta\Phi_J p + \eta_p^J)\right]\cdot\exp(-2\pi i\Phi_J)\cdot J_- \quad (6.156)$$

Also the following term was defined:

$$\Phi = \frac{d}{2\lambda}\cdot\sin(\psi) \quad (6.157)$$

Inserting this into the result of (6.152) gives

$$\text{SIR} \approx \text{SNR}\cdot\left[\pi\frac{d}{\lambda}(\psi_S - \psi_J)\right]^2 \quad (6.158)$$

Using the approximation for the antenna beamwidth, this expression can be compared with the result of the approximation given in (6.35), where D refers to the size of the antenna and d refers to the separation of the two subarrays (U and L)

$$\text{BW} \approx \frac{0.9\lambda}{D} \quad (6.159)$$

The key for this EP algorithm is the high speed of the ASM (v) and the weave maneuver when combined with the optimal digital signal processing of the full data array from the two receivers. Originally, this weave maneuver was implemented to mitigate the effectiveness of kinetic defenses from the naval fleet, and the two receivers were adopted to implement improved monopulse processing.

The stated objective of the original PRC references is to evaluate the degraded performance of STAP when limited to two receivers (Σ and Δ channels) and when near forward looking, compared to the typical SAR application of STAP to side-looking radar on an aircraft platform. While the STAP performance is degraded from the original application it is known that STAP performance improves as the speed of the platform increases. Thus, the PRC research of the degradation of SAR for near-forward looking two-channel STAP is a natural EP application against EA (such as false targets and especially cover jamming) for the ASM application.

Simulation Results

A simple simulation of the above model was developed using MathCad. The number of pulses (P) was set to 16 and the JSR to 30 dB. The ASM speed was set to 550 m/s. Also, in the simulation the HVU ship is a simple point target with a nominal Doppler value. Runs consisted of 10 samples for each of 3 SNR values and at each angular separation between the EA source and the HVU. A separation angle of 0.5° corresponds to a cross-range separation of the ship target and the EA platform of about 175m at a range of 20 km.

The magnitude of Z_1 is the cover noise jamming level in a cell that does not contain the target ship. The metric (SIR) is the difference between the level in the cell containing the ship compared to the many cells that do not contain the ship. Taking the mean of 10 samples the metric was computed (in decibels)

$$\text{metric} = 10 \cdot \log \left[\frac{(|Z_0 - Z_1|)^2}{(|Z_1|)^2} \right] \quad (6.160)$$

The results of the simulation and the theoretical value for the SIR approximated above are plotted in Figure 6.9. The simulation results are in agreement with the theoretical predictions. The scatter in the simulation results is because of only having 10 samples for each data point. Additional runs were made at various other JSR or JNR. As expected the results do not vary with EA levels.

The theoretical results for this EP algorithm are repeated in Figure 6.10. These results can be directly compared with the results for the monopulse approximation results plotted in Figure 6.11. Recall that the SNR when executing detection processing in the Σ channel array is 3 dB less than the corresponding SNR when executing detection signal processing in the full array of data. Therefore, the results in the two figures are plotted for directly comparable receiver SNR data conditions and at the same angular separation between the EA platform and the target ship.

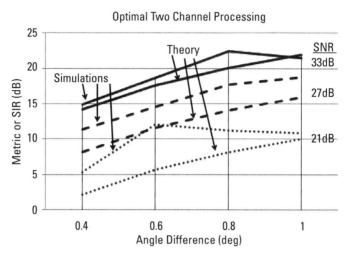

Figure 6.9 EP algorithm simulation results and theoretical values.

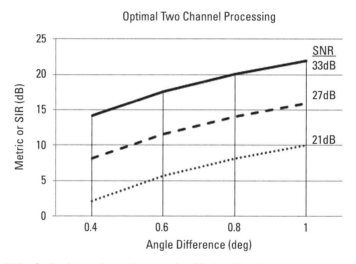

Figure 6.10 Optimal two-channel processing EP algorithm theory.

Comparing the optimal two-channel signal processing results with the monopulse approximation results shows that the optimal processing gives about a 6-dB improvement for these cases. A detailed comparison depends on the implementation details. Half of the improvement shown comes simply from proper use of the full array of data available. The optimal two-channel technique not only has better performance, but is more efficient to implement both in the hardware attributes and in the overall signal processing burden.

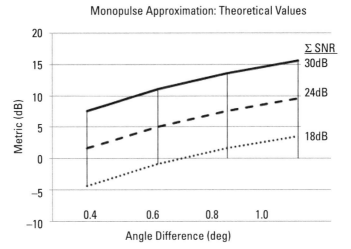

Figure 6.11 Monopulse approximation EP algorithm theory.

As mentioned above, these results can be further improved with additional coherent processing gain (that is, higher SNR from pulse compression gain or additional Doppler gain). Any processing gain can be added to the vertical axis in the Figures 6.10 and 6.11. For example, adding a pulse compression gain of 100 results in a 10-dB increase in the vertical axis values. Or equivalently, if the target corresponds to a lower radar cross-section target such as a newer stealth naval ship, this could easily be offset by including more coherent processing gain.

These techniques have been applied to actual radar simulator captive carry test data and were found to be valid. The hardware simulator tests were conducted primarily with the monopulse approximation technique since the field test conditions could not achieve the typical speeds of the ASM (Mach 1 to Mach 3 or more). That is, the optimal techniques were not hardware field tested since the platform speeds were low (about 250 kts). Testing of the monopulse approximation algorithm has shown that cover EA can be rendered basically transparent to various coherent two-channel processing techniques for the ASM scenario as are being developed by PRC ASM engineers.

6.4 Summary

A simple model has been utilized for the purpose of elucidating several coherent two-receiver EP approaches for mitigating both cover EA and false target

EP. The full expressions for the mathematical physics-based model target echo, EA false targets, EA cover jamming, and receiver noise were presented.

Multiple false targets can be readily identified by various correlation methods detailed in several PRC publications. For the case of the EA platform aligned in angle with the target ship, the PRC techniques identify various antenna polarization techniques to distinguish the true target. Several parameters were shown that correlate the several false targets. Several of these techniques have been tested and verified by the U.S. Navy in radar hardware simulator testing.

A classically very effective method of denying the ASM any information about the target ship (HVU) is the use of cover or noise jamming EA. The only recourse of the ASM processor in the past has been to track in angle only (HOJ) until burn through occurs. If this occurs after significant angular separation has been achieved the ASM is generally defeated.

This classical result is based on the assumption that the ASM processor utilizes only Σ channel information for target detection and identification. A monopulse approximation and an optimal two-receiver processing technique were explored based on publications by PRC personnel. While the Σ channel data is totally corrupted with noise EA the monopulse data presents a uniform and consistent background level in all Doppler range cells except the cell containing the HVU. The cell containing the HVU is easily distinguishable from the background and readily detectable. It is shown that the HVU can be detected, identified, and tracked based on the angular separation and the SNR and is rather insensitive to JSR. The performance metric is approximately (where BW is the ASM antenna beamwidth and SNR refers to the SNR in the Σ channel)

$$|\text{metric}| \approx \left|\left(\psi_S - \psi_J\right)^2\right| \cdot \frac{1.885^2}{\text{BW}^2} \cdot \text{SNR} \qquad (6.161)$$

The alternative approach is to use a prewhitened steering vector and to perform processing analogous to standard STAP, thereby exploiting the full array of data available. This algorithm also exploits the high-speed weave maneuver originally implemented to mitigate the fire control solutions of kinetic defensive weapons. Again, these techniques render cover EA transparent. The performance is (where the SNR is the full two-channel SNR and therefore 3 dB greater than the SNR in the previous formula)

$$\text{SIR} \approx \text{SNR} \cdot \left[2\pi\left(\Phi_S - \Phi_J\right)\right]^2 \qquad (6.162)$$

These techniques were found to be at least 6 dB more effective than the monopulse approximation, and the performance can be significantly improved with additional receiver processing gain. Implementing this class of algorithms also makes possible a much simpler seeker hardware configuration and a more identical use of the parallel radar receivers.

In sum, optimal two-receiver processing results in EP techniques for the ASM coherent radar that basically renders cover EA transparent to the radar processor. These techniques are being actively pursued by PRC and other ASM development engineering personnel.

References

[1] Wang, H., et al., "An Improved and Affordable Space-Time Adaptive Processing Approach," *1996 CIE International Conference of Radar Proceedings*, Bejing, China, October 8–10, 1996, pp. 72–77.

[2] Wang, H., and Y. Zhang, "Further Results of $\Sigma\Delta$-STAP Approach to Airborne Surveillance Radars," *1997 IEEE National Radar Conference*, Syracuse, NY, May 13–15, 1997, pp. 337–342.

[3] Wang, H., "STAP for Clutter Suppression with Sum and Difference Beams," *IEEE Trans. on Aerospace and Electronic Systems*, Vol. 36, No. 2, April 2000, pp. 634–646.

[4] Wang, H., et al., "$\Sigma\Delta$-STAP: An Efficient Affordable Approach for Clutter Suppression," In *Applications of Space-Time Adaptive Processing*, R. Klemm (ed.), London: The Institute of Engineering and Technology, 2004, pp. 123–148.

[5] Guerci, J., *Space-Time Adaptive Processing for Radar*, Norwood, MA: Artech House, 2003.

[6] Genova, J., "Coherent Seeker Guided Antiship Missile Performance Analysis," NRL/FR/5760-05-10,090, Naval Research Laboratory, Washington, DC, January 28, 2005.

7

Adaptive EW

The working example in this book is that of a naval fleet under attack by waves of autonomous, guided anti-ship missiles. Under these conditions, kinetic defensive weapons alone will not suffice to protect the fleet with a high survival rate and an adequate time on station. In Chapter 1, it was shown that the initial decision of the optimal mix of defensive weapons chosen by the fleet warfighter is based on the analysis of the probability of raid annihilation. The required inputs to this analysis include the variety of fleet weapons and threats available as well as the collection of the a priori probability of weapon effectiveness for each threat and weapon combination. These probabilities result from extensive testing and intelligence gathering.

As the battle progresses the a priori probabilities must be replaced by the real-time estimates of a posteriori probabilities of effectiveness. These probabilities address the kill effectiveness of the weapons for the particular engagement of the moment. These factors are required information in the effort to defend the fleet with the optimal probability of survival and the maximum time on station. Any time a weapon is used its effectiveness must be evaluated to determine if alternative actions are required to ensure the maximum probability of survival.

At the same time the autonomous threat ASM must sense the defensive efforts of the fleet and react accordingly. The ASM sensor is activated when information is required to guide the platform to the proper target. The sensor must detect the target and assess the defensive actions employed by the fleet. The ASM sensor must select the optimal waveforms and processing to quickly and accurately determine which return represents the desired target. All of these tactical decisions (for both the fleet warfighter and the autonomous threat missile) must be computer aided or automatic because of the amount of information to be absorbed and the rapidity of events of the battle.

Designing the defensive weapons (both kinetic and nonkinetic), calculating the probabilities of effectiveness, and defining the battle tactics require knowledge of the ASM sensor capabilities and an understanding of the ASM sensor signal processing algorithms. The ASM radar sensor, using modern technologies and LPI techniques, can readily detect and track the naval target. The modern EW battle is an information battle focused primarily on the task of target identification or target classification [1].

While in the past the goal of EA was to exploit known flaws in the ASM seeker sensor, the goal of modern EA (the nonkinetic weapons) is to counter the ASM sensor task of classifying the target. This information battle must be based on general physical properties rather than specific hardware flaws in the ASM. In particular, the goal is to deny the ASM sensor required target classification feature information and/or to present the ASM sensor with realistic target features associated with one or more false targets or decoys. In its endeavor to choose the correct target the ASM sensor employs rapid and efficient signal processing algorithms. In the language of EW a variety of EP algorithms are employed for the purpose of protecting the ASM sensor from fleet EA techniques.

In the chapters of this book, several ASM sensor EP techniques useful to properly classify a ship target in the presence of EA have been described. In describing these EP signal processing techniques the philosophical direction of EW as a battle for information becomes clearer. In addition, several potential modern EA counters have been indicated as appropriate. A simple model of the basic characteristics of an ASM radar sensor was developed in Chapters 2 and 3, for use in illustrating several novel EP algorithms. The model makes it possible to quantitatively understand the signal processing of EP algorithms. The model enables the engineer to develop an intuitive understanding of the DSP algorithms.

In Chapter 4, it was shown how the ASM sensor exploits the RCS level and several statistics of the RCS measurement, including the mean, the variance, and the correlation function. The EP techniques include the exploitation of the statistics of the monopulse ratio to glean valuable features of the target RCS level. The RCS characteristics are a manifestation of the HVU, being an extended target composed of multiple scatter elements. Standard EA techniques including DRFM-based active EA as well as passive reflector decoys and chaff are all readily distinguishable by the modern ASM sensor processor. The EA techniques generally do not properly replicate the requisite target features that are quickly measured by the ASM sensor. It was shown how the fleet EA system can exploit dual coherent systems such as cross-polarization jamming to mimic several of these characteristics of a complex extended reflector like a naval ship.

In Chapter 5, several novel LPI waveforms made feasible by modern RF technology were shown to be fully capable of providing the standard ASM guidance measurements while at the same time enhancing the EP capability of the ASM. It is especially noteworthy that some of these waveforms are deliberately designed to enhance classification features of the HVU. In a sense the waveforms are designed to probe the targets for classification information, thereby enhancing the EP features and performance of the radar sensor. This again illustrates the shift away from exploiting hardware characteristics and toward deliberately probing the target to stimulate or enhance distinguishing features for target classification.

The existing modern ASM already contains multiple coherent sensor channels for monopulse measurement processing. And the modern ASM already flies a weave trajectory, including high-g turns to mitigate the performance of the fire control algorithms of kinetic weapons. In Chapter 6, it was shown that these existing capabilities make possible the implementation of optimal multiple channel signal processing algorithms. These optimal DSP algorithms perform better than present algorithms by properly combining all of the existing data, thereby doubling the amount of information processed. Fully exploiting the two-receiver channel data already available to the modern ASM makes feasible the improved mitigation of false targets, and also the much improved mitigation of cover noise jamming. Basically, noise jamming is rendered transparent to modern ASM processing. Pointedly, this much improved ASM sensor performance can be implemented in existing systems via a software upgrade. The only viable EA remaining is to create truly representative false targets that can properly seduce the ASM sensor away from the HVU based on true physical principles.

Some or all of these EP techniques will be operated simultaneously during the engagement. This requires the ASM processor to combine multiple different parameters or pieces of information in order to make rapid and automatic combat tactical decisions. To properly understand this information and to properly weight the mix of these measurement techniques is again based on mathematical probabilities and statistics.

It is imperative to recognize that both of these tasks (selection and operation of fleet defensive assets and selection and attack of the proper target by the ASM) are hypotheses testing tasks and require an understanding of signal processing. In this chapter, the fundamentals are described of an algorithm designed to rapidly combine multiple differing measurement parameters to aid the warfighter tactical decision-making tasks and to aid the autonomous weapon tactical decision-making tasks.

Considerable work has been done in the development of algorithms to combine multiple parameters for the purpose of computer-aided decision making. The goal of this chapter is only to indicate the desired properties of the algorithm as applied to the task of the modern EW information battle. To accomplish this goal, the characteristics of a viable but simple mathematical approach to decision making based on a variety of measurements is generally described.

In the first section, the basic terminology and characteristics of the algorithm are explained. The general flow of the algorithm is presented as the fundamentals of the EW battle are described. It is shown how the algorithm measurements and assessments are made feasible and enhanced by knowledge of the tactics being employed. The measurement process must be strongly coupled to knowledge of the tactical situation. Particular detailed understanding of the engagement is not the result of general measurements. Rather, the real-time assessments are the result of detailed understanding of the interplay between tactics and measurements.

In the second section, the mathematics of the basic LLR algorithm is reviewed. This concept has been employed throughout this work. In this way, the properties of the rapid decision-making algorithm are illustrated and examined. In particular, the value of casting the decisions into hypothesis space is manifested by the LLR, being a scalar quantity with known statistics.

In Section 7.3, the specific application to several simple EW examples is explained to enable an examination of the processing algorithm. It is shown that the algorithm is staged in parallel with the stages of the engagement. A significant improvement results from building on prior knowledge rather than from relying on definitive and absolute measurements. The modern RF technology combined with high-speed signal processors and algorithms makes it possible for the ASM sensor to rapidly adapt to the information needs as the engagement evolves. The sensor is becoming much smarter. Several present-day examples of smart weapon system engagements are described for illustrative purposes.

Section 7.4 contains a summary of the fundamental characteristics and results. It is again emphasized that the modern ASM presently has a very significant advantage over the defenses of a naval fleet.

7.1 Overview

For this work, three basic defensive EA actions of ship defense against an ASM are considered for the purpose of providing concrete examples of the theory. Strategy 1 consists of the HVU and a single decoy. This is the most basic strategy. The decoy may be a combination of passive reflectors or an active repeater, or chaff. The EA goal is to capture the ASM tracking gates on the false off-board target as early in the engagement as the ASM seeker does its detection, classification, and localization tasks. The decoy must have features representative of a true target. It is expected that the probability that the ASM selects the HVU is at least 0.5, based on its having prior information and its ability to detect both possible targets.

Strategy 2 consists of the HVU and an off-board platform using some form of seduction EA to defend the HVU. At the onset of Strategy 2 the ASM seeker may or may not be tracking the ship target. The EA consists of some form of electronically generated false target or targets. The EA generated false targets are used in an attempt to capture the tracking gates and to seduce these tracking gates from the ship and ultimately onto an off-board false target. The false targets must have features representative of a true target. The false targets may be collocated with the ship target. In this case the false target must move to another Doppler range location in a nonphysical manner. Or the region can be made to include many false targets to add confusion to the ASM sensor classification task. The goal of the EA in this case is to confuse the ASM target classification and to enhance the probability of the ASM seeker eventually selecting the off-board decoy sufficiently in advance of ship impact.

All fleet defensive strategies should include a decoy as the desired target for the ASM as in Strategy 1. The eventual goal is to reduce the probability that the ASM selects the HVU by either enhancing the probability it selects the decoy or by decreasing the probability that the ASM selects the HVU.

Strategy 3 is to cover or hide the true targets with noise jamming. If an off-board device generates cover noise jamming, it can seek to seduce the ASM by concealing all targets during the early portion of the terminal phase of the attack and putting the ASM in home on jam (HOJ or angle track only). Obviously the goal is to greatly decrease the probability that the ASM will select the HVU. If cover jamming is generated by a ship, then it must be combined with some additional means such as Strategy 2 to seduce the ASM to switch to an off-board device later in the engagement as in Strategy 1. Thus, Strategy 3 includes the option of the HVU using onboard EA for self-protection as in Strategy 2 and/or the use of a decoy as in Strategy 1.

The working example in this book is that of a naval fleet under attack by waves of autonomous, guided ASMs. All of the tactical decisions (for both the fleet warfighter and the autonomous threat missile) must be computer aided or automatic because of the amount of information to be absorbed and the rapidity of events of the battle. A functional model of the decision process (for either the fleet warfighter or the autonomous ASM) is presented in Figure 7.1.

Data is collected by one or more sensors. Based on these observations, a real-time model of the engagement is developed by the observer algorithm. From the fleet perspective, tactical weapons actions and various sensor actions and modifications are executed at the command of the controller algorithm based on the estimate in the observer of the real-time engagement situation. From the ASM perspective, its observer assesses the feasibility of each of the targets to be a true target and assesses the possible EA actions being employed. Based on this assessment of the observer, guidance commands are executed at the command of the controller algorithm. In addition, the assessment of the observer may lead the controller to direct the sensors to collect specific information to improve its ability to assess the engagement situation. It is important to note that the performance of the observer is greatly enhanced

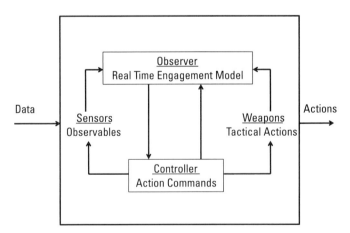

Figure 7.1 Decision process.

by accumulating information and by knowledge of the controller actions in both the cases.

At the onset of each part of the battle the choice of the mix of defensive weapons and tactics chosen by the fleet warfighter is based on the analysis of the probability of raid annihilation. This analysis depends on the a priori probabilities of kill based on intelligence data and previous system testing. These probabilities are only as good as the intelligence data and realism and fidelity of the testing.

Any time a weapon is used, its real-time effectiveness must be continually evaluated to determine if alternative actions are required to ensure the maximum probability of survival. When kinetic weapons are used, fleet sensors are employed to monitor the ASM for physical damage or significant trajectory changes caused by physical damage coincident with impact.

The tactical use of real-time EA effectiveness monitoring (RTEAM) was developed for the U.S. Navy in the 1980s during several programs and has been previously thoroughly tested. The results of one of these efforts are presented herein as an example of the philosophy of modern EW.

The threats (ASM) are described by their basic tactical characteristics. The threats execute events along a preferred engagement timeline in the effort to guide the threat to impact with a target ship (using the goal: miss distance < the critical miss distance). The threat engagement timeline is heuristically represented in Figure 7.2.

As an example of a portion of such a sequence the ASM may fly to a particular geographic position and pop-up over the radar horizon. The seeker is turned on and does an antenna sweep to collect data over a defined range swath and angular direction where a ship of known characteristics had been previously seen. Once the possible ship location is verified and updated the ASM moves to the next node on its engagement timeline.

At this node, the sensor may be directed to probe for additional target features in an effort to classify potential targets as the desired target, or as not the desired target. Once the target selection is confirmed the ASM moves to the track mode and its final attack phase. If the target is assessed to be a false target an alternative target is chosen.

Each decision node on the ASM engagement timeline is described by its characteristics, including its vulnerability as an event that can be exploited via an action by the variety of weapons available to the fleet. The goal of the fleet weapon action is to cause the threat to transition from the ASM desired engagement timeline that leads to target impact to an engagement timeline desired by the defending forces that lead to the threat not impacting a target ship as indicated by the modified Figure 7.3.

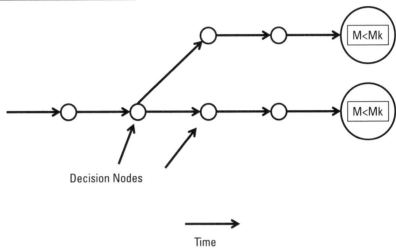

M = Miss Distance Estimate
Mk = Design Goal

Figure 7.2 Threat engagement timeline.

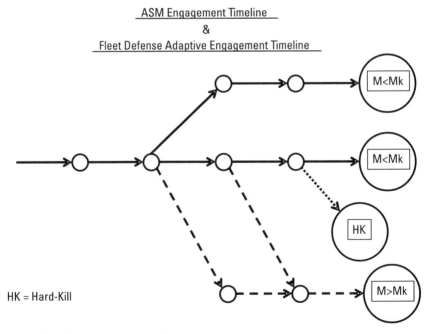

HK = Hard-Kill

Figure 7.3 Altered engagement timeline.

The above description can be symbolized as shown in Figure 7.4. The threat (ASM) is described by its engagement timeline. Each node is further described by its vulnerabilities to particular weapons and weapon actions. Any time a weapon action is executed, there is a particular goal desired. This general goal is to prevent the threat from continuing on the ASM engagement timeline and to have the ASM transition to a fleet desired timeline. At each node that is attacked, there is the possibility that the ASM continues engaging the ship (H0) or not engaging the ship (H1). These hypothesis options must be characterized by any and all observables that are possible by the fleet sensors. Thus the weapon controller directs the sensors to collect the appropriate data for the observer.

In the same way, the autonomous ASM can view the battle engagement symbolically. The ASM has a desired engagement timeline. There are decision nodes on its timeline where it must direct its sensor or sensors to gather data to evaluate its progression of the engagement. At each node the ASM must evaluate the correctness of its decisions and the effectiveness of fleet actions as well as ASM actions.

Applying this concept to EW, it is necessary to implement some level of real-time effectiveness assessment. If a weapon's effectiveness cannot be evaluated, it is not viable to rely on this weapon except as a last resort. Present fleet warfighter doctrine primarily relies on kinetic weapons. The reason for this is that the real-time effectiveness of kinetic weapons can be evaluated for the warfighter.

To recommend that the warfighter rely on the perhaps more cost-effective EW weapon requires RTEAM. The gist of RTEAM is to evaluate the decision made at a particular decision node to determine the resultant timeline decisions. In the simple case the outcome of the decision node, given a known

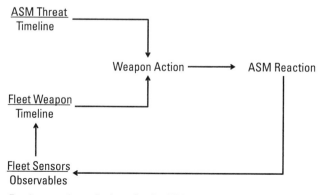

Figure 7.4 Fundamental terminology for the EW battle.

action by the weapon system, is the hypothesis option EA desired engagement timeline or the alternate hypothesis option threat desired engagement timeline.

RTEAM algorithms may be understood by the example of classical hypothesis testing. This chapter summarizes the application of LLR math to EA hypothesis testing. Other mathematical formalisms may be found to be more useful. Regardless, this formalism will illustrate the several fundamental aspects of RTEAM.

As an example, assume the following sequence as a very simplistic engagement time-line (Figure 7.5) (This sequence is from the Advanced Technology Demonstration [ATD] program conducted with the Naval Research Laboratory in 1990, based on research developments performed from 1978 to 1990.)

Engagement Sequence from the Fleet Perspective

1. A priori data is fed to the threat and it is launched.
2. At approximately 12 nmi (range to go) the RF seeker does a Search; Search observed.
3. By 10 nmi the threat has successful detection. The seeker transitions to Track; Track Mode on ship observed.
4. At approximately 8 nmi the onboard EA executes cover EA with a keeper pulse (keeper is 0.25 nmi beyond the ship; the keeper pulse and the decoy are at the same range).

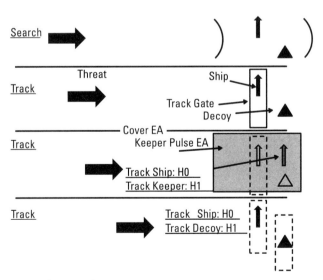

Figure 7.5 Simplistic scenario.

5. At approximately 6-nmi, EA effectiveness is evaluated. Threat tracking ship (H0) or Keeper (H1). If ASM tracking keeper pulse the onboard EA is discontinued.
6. The threat continues Track of either the ship (H0) or the decoy (H1).
7. Effectiveness is estimated. If H1, no action. If H0, hard-kill employed.

At the same time that the fleet warfighter is making these tactical decisions for fleet defense, the ASM is executing its engagement timeline and making decisions to improve its effectiveness (probability of kill). The ASM flies to a preselected location and commands the seeker to update target information. Having detected the target possibilities, the signal processor evaluates its data to select the object most representative of the predefined target.

The ASM must decide via hypothesis testing algorithms which object to choose as its preferred target. Once the ASM selects a target, it makes measurements for guidance and it continually makes measurements to develop tactical decisions in an effort to protect itself from attack by EA or kinetic weapons.

For the engagement illustrated in Figure 7.5, the ASM selects the correct target. At this stage the ASM detects cover jamming. It may switch to HOJ mode or it may detect the keeper pulse as a viable target. When the jamming is discontinued the true target and the off-board decoy are in the ASM sensor view. The ASM may be in track mode and continue to track the decoy since it is already in the tracking gates. The ASM may identify the decoy as a false target and return to tracking the ship. In either case, the ASM must determine if it has time to assess the fidelity of the targets and alter its course accordingly.

Engagement Sequence from the ASM Perspective

1. A priori data is fed to the threat and it is launched.
2. At approximately 12 nmi (range to go) the RF seeker does a search for previously defined target.
3. By 10 nmi the threat has successful detection of two objects. The seeker evaluates target features and transitions to track mode on ship.
4. At approximately 8 nmi the onboard EA executes cover EA with a keeper pulse (keeper is 0.25 nmi beyond the ship; the keeper pulse and the decoy are at the same range). ASM executes track mode on keeper and evaluates target features or switches to HOJ based on prior observations of ship target.
5. At approximately 6 nmi the onboard EA is discontinued. Target features are evaluated. ASM commits to ship or decoy.

In the previous chapters, several ASM sensor EP techniques useful to properly classify a ship target in the presence of EA have been described. In describing these EP signal processing techniques the philosophical direction of EW as a battle for information becomes clearer.

The primary goal of modern EA (the nonkinetic weapons) is to counter the ASM sensor task of classifying the target. In particular, the goal is to deny the ASM sensor required target classification feature information and/or to present the ASM sensor with realistic target features of one or more false targets or decoys. In its endeavor to choose the correct target the ASM sensor employs rapid and efficient signal processing algorithms. In the language of EW, a variety of EP algorithms are being developed for the purpose of protecting the ASM sensor from fleet target deceptive and denial jamming techniques.

Considerable work has been done over the years in the development of algorithms to combine multiple parameters for the purpose of computer-aided decision making. When the task is to quickly select the correct object from many possibilities, one standard approach is to use a staged probability estimate algorithm. The first stage has poor statistics but quickly reduces the number of options while including the correct object with high certainty. The probability of a false alarm (PFA) is set relatively high with a correspondingly very high probability of detection. In this way the true target is definitely accepted but false targets may also be selected as possibilities.

This stage is generally very processing efficient and very fast. The objective of this stage is to quickly reduce the number of choices of interest. This stage is typically employed by the ASM in its search mode or target reacquisition mode. In this manner many potential targets, including several ships, decoys, and false targets, may be detected in the range swath as the antenna is swept through some angular swath. Target probabilities are weighted based on previous data. The modern ASM seeker sensor combined with multiple DSPs has the capability to process multiple target features from multiple targets simultaneously.

This stage is followed by a mode that has finer PFA, but is more computationally intensive. This mode is typically the ASM algorithm for target selection or classification of the multiple target possibilities identified in the search mode. This mode can include the measurement of one or more target features. All of this information must be combined to estimate which of the possibilities is the most likely target of interest. Once the determination has been made to identify the correct target the ASM switches to its target track mode. During this final mode, localization measurements are made as the ASM autonomously guides to target impact.

As mentioned above, between 1978 and 1990 the author developed an algorithm to combine a variety of diverse data for the purpose of making real-time tactical decisions. This program culminated with an ATD onboard a U.S. Navy ship. This algorithm using classical LLR methodology was implemented and tested for the fleet defense task, but is equally applicable to the task of automatically mixing data for the ASM EP task. This algorithm has many of the properties desired of the ASM algorithm to combine target feature data. The desired properties of such an algorithm are described below.

7.2 Fundamentals of LLR

In this section, a simplistic LLR formalism is described. The goal of this example is to demonstrate some of the requirements of a tactical decision aid algorithm. For example, this algorithm may be used to evaluate the state as defined in step 6 of the fleet defense sequence illustrated above. As described above, an action by an EW weapon is executed for the purpose of controlling the threat to a preferred engagement timeline. The impact of this action on one or more sensors (observables) is examined for the purpose of evaluating the result of the action such as indicated in step 6 above. Or a decision aid algorithm may be used to evaluate the possible targets in step 3 or step 5 of the ASM tactical decisions. In this case, the LLR formalism is used to combine the several target features into a single decision parameter.

Assume a sensor provides a single measurement or a time sequence of N measurements (1s or less) of data represented by x. If hypothesis H_i is true the data is represented as the following set of measurements:

$$Hi = \left\{ x | x_n = \mu_i + \delta x_i \frac{n}{N-1} + \omega_n \right\} \quad (7.1)$$

$$N \text{ is odd}; \; n = \left[-\frac{N-1}{2}, \frac{N-1}{2} \right] \quad (7.2)$$

The μ parameters represent the means. The δx parameters represent the span of the data over the time interval, and ω represents additive noise for each data sample. An example of the two data sets is illustrated in Figure 7.6.

The solid curves represent the data sets (without noise) with the key parameters indicated for the two hypotheses. The parameters illustrate the

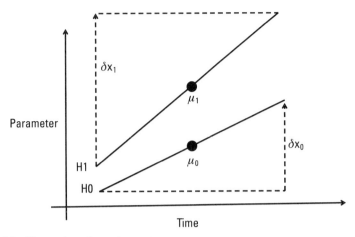

Figure 7.6 Illustration of two data sets.

mean value and the extent of the values, assuming a linear progression with time (or first-order Taylor approximation). This approximation is generally valid since the total time available is short. If $N = 1$ the second parameter is not included. The time interval (T) is related to the sample rate

$$T = N \cdot \delta t = N \cdot \mathrm{SR}^{-1} \tag{7.3}$$

Assume that the sensor measurement noise (ω_n) is Gaussian and uncorrelated for each time sample. The measurement noise variance is σ^2. LLR (λ) is defined as the ratio of the conditional probabilities.

$$\lambda = \log\left[\frac{P(x|\mathrm{H0})}{P(x|\mathrm{H1})}\right] \tag{7.4}$$

With this definition the measurements data set is converted to a set of scalar quantities. Since the noise samples are independent and using the assumption of a Gaussian distribution, LLR can be expressed by the following set of equations and terms. (It is straightforward to modify these expressions for the case of $N = 1$.)

$$\lambda = T \cdot \mathrm{SR} \cdot \left[(\bar{x} - \bar{\mu}) \cdot \left(\frac{\mu_0 - \mu_1}{\sigma^2}\right) + (\Delta x - \bar{\delta}) \cdot \left(\frac{\delta x_0 - \delta x_1}{12 \cdot \sigma^2}\right)\right] \tag{7.5}$$

$$\bar{x} = \frac{1}{N}\sum x_n \qquad (7.6)$$

$$\Delta x = \frac{12}{N\cdot(N+1)}\sum nx_n \qquad (7.7)$$

$$\bar{\mu} = \frac{\mu_0 + \mu_1}{2} \qquad (7.8)$$

$$\bar{\delta} = \frac{\delta x_0 + \delta x_1}{2} \qquad (7.9)$$

Now, as in previous chapters, define the quantity

$$\Lambda = T\cdot\text{SR}\cdot\left[\frac{(\mu_0 - \mu_1)^2}{\sigma^2} + \frac{(\delta x_0 - \delta x_1)^2}{\sigma^2}\right] \qquad (7.10)$$

This quantity is a measure of the separation of the two data sets compared to the sensor measurement accuracy (σ^2/N) (see [7.3]). From the assumption of the Gaussian probability distribution this quantity fully defines the probability distributions for the LLR for the two hypotheses

$$\text{mean}(\lambda|\text{H0}) = -\text{mean}(\lambda|\text{H1}) = \frac{\Lambda}{2} \qquad (7.11)$$

$$\text{var}(\lambda|\text{H0}) = \text{var}(\lambda|\text{H1}) = \Lambda \qquad (7.12)$$

Thus, the distributions overlap a great deal for small Λ and separate as Λ increases. That is, the width of the distributions is the square root of Λ, and the separation of the means of the distributions is Λ. Sample probability distributions (H1 on the left and H0 on the right) for increasing Λ are shown in Figure 7.7.

For a given test performance criteria, such as the Neyman-Pearson criteria, there is a critical Λ value. Thus, as in Figure 7.8, a threshold Th is set such that the probability of accepting H0 when H1 is true is given as b, the area under the dashed curve to the right of Th. As Λ increases there comes a critical value such that the probability of rejecting H0 when it is true reaches the desired performance value as a where a is the area under the solid curve to the left of Th. For example, if $b = 0.0001$ and $a = 0.1$ the critical value is

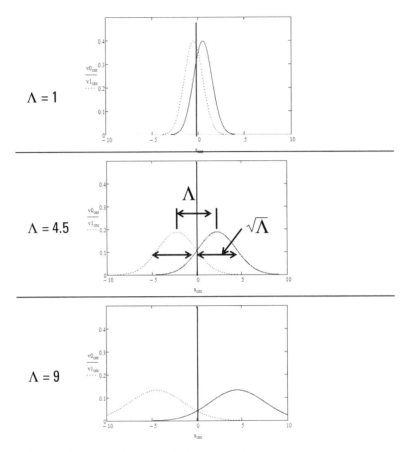

Figure 7.7 Probability distributions for increasing Λ.

$\Lambda = 25$. As Λ continues to increase the threshold is set by the b requirement and the test trivially satisfies the performance requirements.

Typically for the EW scenario application, the value of Λ increases as the engagement range decreases. Therefore, the critical LLR variance corresponds to a critical engagement range. That is, the decisions required can be made with the desired fidelity at a critical engagement range or less. The performance goal can be equated to an engagement range criterion.

Reading the performance threshold values (T_a and T_b) for a standard normal distribution, it is seen that adequate decision performance corresponds to

$$Th = \sqrt{\Lambda} \cdot \frac{T_b + T_a}{2} \qquad (7.13)$$

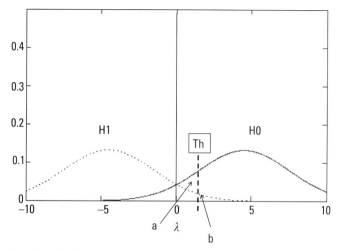

Figure 7.8 Decision threshold.

$$\sqrt{\Lambda} \geq (T_b - T_a) = \sqrt{\Lambda_{critical}} \qquad (7.14)$$

The performance can also be analyzed via the standard receiver operating characteristics (ROC) analysis. These curves are sometimes designated detector operating characteristics (DOC). The *x*-axis is the probability of choosing H0 when H1 is true (or parameter b). The *y*-axis is the probability of correctly choosing H0 when H0 is true (or the parameter 1 − a). As shown in Figure 7.9 the line for $\Lambda = 0$ is the diagonal. When Λ is zero the two distributions are identical. As Λ increases the curve bows more and more to the upper left.

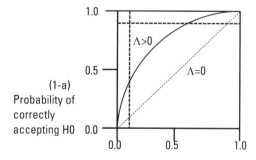

Probability of incorrectly accepting H0 (b)

Figure 7.9 ROC curve representation.

The Neyman-Pearson performance criteria are represented as vertical and horizontal dashed lines. The vertical dashed line indicates the error type *b* (probability of accepting H0 when H1 is true). The horizontal dashed line indicates the error type *a* (probability of rejecting H0 when H0 is true). Acceptable performance corresponds to curves to the left of the vertical dashed line. As Λ increases the curve reaches the intersection of the horizontal and vertical dashed lines (at the critical Λ). As Λ increases the performance requirements are exceeded. ROC or equivalently DOC analysis has been well developed over the years. All of the standard ROC chart analysis can be utilized for LLR.

As another way to view the results, consider again the performance parameter Λ, basically (7.10). Consider the case of a single sensor that is described by its accuracy (σ) and its data rate (sample rate). Figure 7.10 demonstrates an analysis of the situation. A graph of sensor characteristics can be created by letting the vertical axis be the sensor accuracy and the horizontal axis the sample rate. Increasing along the vertical axis represents a less accurate sensor. Increasing along the horizontal axis represents a higher sample rate sensor.

The solid parabolic curve represents the critical performance value of Λ for a time fixed situation or fixed geometry scenario. Any higher value of Λ will meet the performance criteria. This curve corresponds to the hypothesis geometry values in (7.10) being fixed. The acceptable operating region is the clear region to the right and under this Λ curve.

The vertical solid line represents the maximum sample rate achievable. It is assumed that sensors may be available at this sample rate or less. The horizontal solid line represents the best accuracy achievable (smallest σ).

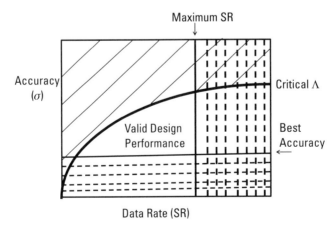

Figure 7.10 Sensor operating requirements.

Again less accurate sensors may be available. An available sensor can assess the effectiveness of this EA action when its specifications result in the clear area as shown in the figure.

As stated above the critical performance parameter is generally a function of engagement range, and this parameter generally increases as the engagement range decreases. Therefore, the critical value of Λ for a single sensor corresponds to a critical engagement range as illustrated in Figure 7.11 that shows the performance parameter plotted versus engagement range.

A major issue for many tactical situations is the task of sensor or data fusion. In many cases, this involves complex and processor-intensive coordinate transformations and algorithms. Tactical decision making must be rapid and efficient. The method of sensor fusion for decision making is a major advantage of this LLR formalism. Consider the results of applying the formalism to two independent sensors (A and B) for the same event. If the sensors are independent, the resultant LLR formed by combining the LLR is Gaussian. The resultant LLR is simply the sum of the two independent LLRs, and the variance of the resultant probability distribution is the sum of the individual sensor variances.

$$\lambda = \lambda_A + \lambda_B \tag{7.15}$$

$$\Lambda = \Lambda_A + \Lambda_B \tag{7.16}$$

That is, sensor fusion is achieved by adding the scalar LLR values and the performance is described by exploiting the addition of the variances. As stated above, the decision test performance improves with increasing Λ. The sum of

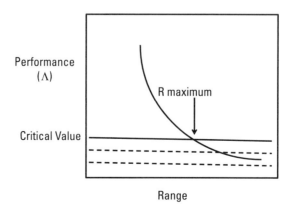

Figure 7.11 Performance versus range.

the two variances must be greater than either individual variance value. The relative weighting of the two LLRs in how they impact the overall decision is automatic, based on their variances, and the test performance is improved with this larger variance.

Again, sensor fusion is achieved by scalar addition with the proper weighting. There is no need to justify coordinate systems and so forth, to achieve optimal sensor fusion. In the same manner, independent time samples can be added as scalars. However, depending on the tactical situation, it may be advisable to apply a fading memory filter to this time information or to make new measurements if the situation may have changed. The performance improvement resulting from sensor fusion is illustrated in Figure 7.12.

The main features of the LLR type of decision algorithm are as follows. Typically, an action occurs that will leave the situation in different possible states. The required decision is to evaluate which state is most likely. One or more observable quantities are measured. The measurement is modeled. Since the decisions must be made rapidly, only short time intervals (or individual measurements) are involved. A simple approximation to the measurement variable can be made. The sensor parameters such as accuracy and sample rate are modeled. Assuming Gaussian probability distributions the LLR is formed. If there are more than one measurement variable the individual LLRs are combined. If the test performance is viable (Λ larger than the critical value) the measured value of the LLR is compared with a threshold value.

Figure 7.13 provides a simplified geometry of the engagement and defines the several variables. The goal of the fleet-based algorithm is to measure the a posteriori probability of EA effectiveness. This may result from an estimate of the miss distance. A miss distance of M indicates the ASM is targeting the decoy. A miss distance less than M indicates the ASM is targeting the ship.

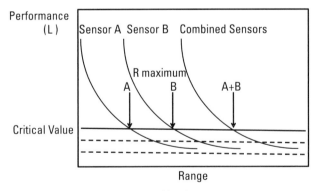

Figure 7.12 Maximum test range using combined sensors.

If the ASM does not maneuver this estimate may be adequate. If the ASM maneuvers this estimate is seldom useful.

Equivalently, EA effectiveness may result from an estimate of the ASM seeker look direction in its final terminal phase. If the antenna is pointed at the ship it is assumed the ASM is targeting the ship. If the antenna is pointed at the decoy then the ship is at angle β relative to the ASM antenna bore sight. This is a much more difficult parameter to estimate, but it is much more reliable in estimating the state of the ASM and the effectiveness of fleet EA as long as a mechanically steered antenna is employed.

From the fleet perspective, the various LLRs can be seen in Figure 7.14. Using the simple geometry in Figure 7.13 the approximation equation (7.5) was computed and these results applied to (7.10). Basic observables examined include bearing (sensor to threat), range, power, polarization ratio, and range rate. Some typical results are shown in the chart in Figure 7.14. All of these variables demonstrate that the test performance improves with decreasing range and/or increasing decoy miss distance, as expected. Using these results the sensor specifications can be determined to allow RTEAM decisions when an EW strategy is defined as above.

Now consider a simple case of a sea-skimming and nonmaneuvering threat (two-dimensional geometry). As stated above, some aspects are estimated over time. The most reliable and useful parameters are associated with a change of state actions. Assume the ASM is in a particular state such as tracking the ship. If the EA action is employed with the goal of changing the state to tracking the decoy, several parameters may be monitored to assess effectiveness of the action. If the parameter does not change, the EA did not change the ASM state. If the parameter has a change coincident with the EA action

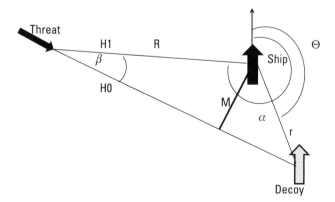

Figure 7.13 Miss distance geometry.

	Bearing (a)	Range (R)	Range Rate dR/dt	Power (A/[R*R])
H1 (Hit ship)	α	$R0 - vn/SR$	$-v$	$P0 - P0 \cdot \left(\frac{2vn}{R0 \cdot SR}\right)$
H0 (Miss Ship)	$\alpha - \frac{vn}{SR} \cdot \left(\frac{M^2}{R0^2}\right)$	$R0 - \frac{vn}{SR} \cdot \left(1 - \frac{M^2}{2R0^2}\right)$	$-v \cdot \left(1 - \frac{M^2}{2R0^2}\right)$	$P0 - P0 \cdot \left(\frac{2vn}{R0 \cdot SR}\right) - P0 \cdot \left(\frac{M^2}{R0^2}\right) \cdot \left(\frac{vn}{R0 \cdot SR}\right)$
Λ	$\frac{T \cdot SR}{\sigma^2} \cdot \left[\left(\frac{M}{R0^2}\right) \cdot vT\right]^2$ $\frac{T \cdot SR}{\sigma^2} \cdot \left[\left(\frac{M^2}{2R0^2}\right) \cdot vT\right]^2$		$\frac{T \cdot SR}{\sigma^2} \cdot \left[\left(\frac{vTM^2}{2R0^2}\right)\right]^2$	$\frac{T \cdot SR}{\sigma^2} \cdot \left[P0 \cdot \left(\frac{M^2}{2R0^2}\right) \cdot vT/R0\right]^2$

Figure 7.14 Table of sample results from fleet perspective.

the EA is assessed to have successfully effected a change of state. In previous testing the time for desired transition from Track ship to Track decoy was known. The goal was to observe a change in the particular observable value consistent with this transition of the engagement timeline. Sample results are shown in Figure 7.15.

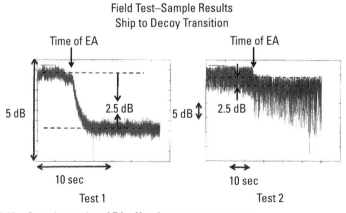

Figure 7.15 Sample results of EA effectiveness assessment.

In a similar manner the variety of parameters described in the previous chapters are monitored and measured by the ASM seeker sensor and digital signal processor. These several parameters are converted to a scalar parameter, such as the LLR, and are combined. In this manner the ASM can rapidly and effectively assess the several targets for physics-based features to optimally determine the correct target from among the several possible targets that are readily detected and tracked. For example, the DSP can monitor the target RCS value, the RCS Lag-1 parameter, and the Doppler dimension of the targets simultaneously. These parameter values can be evaluated via the LLR formalism or in a multidimensional feature space to determine the most probable true ship target. Figure 7.16 contains a list of some of the EP parameters available to the modern ASM sensor that have been described in the previous chapters.

In the next section, some concluding comments are provided for EW-specific examples.

7.3 EW Specifics

Throughout history and especially during and since the Second World War, naval forces have been a means to project power into areas with concentrated hostile forces. Arguably, the superiority of the British fleet in WWII was a

Sample of EP parameters available to the modern ASM sensor
RCS mean
RCS variance
RCS Lag-1
Monopulse variance
BFD
Doppler ghost images
Probed Doppler smear
Correlated RCS fluctuations
Correlated monopulse errors
Target range when using long random code
Target range length/denseness/sparseness
Target Doppler length/denseness/sparseness
Doppler image variation synchronized with weave maneuver
Monopulse approximation to mitigate cover EA
$\Delta\Sigma$ STAP processing to correlate multiple false targets
$\Delta\Sigma$ STAP processing to mitigate cover EA

Figure 7.16 List of sample EP parameters.

main contributing factor in enabling the Allied nations to survive against the Axis powers in the early years of that conflict.

Since that conflict, naval forces have continued to serve at various times to project power into many areas of the world from the Falklands War to the several conflicts in the Middle East and into areas of Asia such as Vietnam, Korea, and Taiwan.

Beginning with Japanese suicide air attacks on naval vessels and the early German development of ASMs, the ASM has been continually developed by several nations as a counter to these naval forces. A simple search of the internet or a comprehensive text [1, 2] reveals that there is a growing arsenal of ASM deployed by several nations, including the major powers (e.g., the United States, China, Russia, and India) as well as lesser military powers (e.g., Iran, Korea, and Pakistan).

While there have been limited battles since WWII as described in Chapter 1, the potential for a major attack on naval forces is on the horizon. At the present time the standard defense for the naval fleet when attacked by the ASM threat is with a layered defense of kinetic weapons. This defense begins with longer range to intermediate range antimissile missiles and concludes with close in weapons system or a laser beam gun.

While this defense may be adequate when confronted with a limited attack by a less capable military power, the kinetic weapons will be overwhelmed when confronted with a major simultaneous attack by modern ASM threats. The potential is ever-increasing of an attack against a naval fleet by waves of ASM of a variety of types.

For example, consider a hypothetical attack against a U.S. fleet by People's Republic of China forces. First, the fleet can be continuously monitored by the array of over the horizon radars and the data collection from reconnaissance satellites. With this targeting information, waves of DF-21 ASMs launched from a variety of platforms will approach from on high at Mach 10, generally targeting the HVUs. At the same time, waves of low-flying cruise ASMs launched from land, air, surface, and subsurface platforms can approach at Mach 1 from a variety of directions. These ASMs may be targeting the HVUs and some may be deliberately targeting escort ships. On final approach, these threats can accelerate to Mach 3 or more as they are automatically guided to the targets. Most if not all of these threats will perform high-g maneuvers in an effort to mitigate kinetic weapon fire control systems.

To defend against such an attack, naval forces have been developing a defensive mix of kinetic and nonkinetic weapons. The premise of this work is that these present autonomous ASMs are equipped with one or more sensors of unprecedented capability combined with sophisticated signal processing capability. In the past, the EW battle paradigm was to exploit flaws in the sensors identified through intelligence gathering. The present EW battle must now be a battle for information that is physics based.

The ASM equipped with LPI radar is very difficult to detect. These radars may be in common weather radar frequency bands or at higher frequency. On pop-up and reacquisition at the onset of the terminal phase the sprint vehicle begins to conduct evasive maneuvers at high speed to mitigate kinetic weapon fire control, while the bus platform presents a decoy to further confound kinetic fire control systems.

The ASM seeker can readily detect and track several targets. The EW battle becomes an information battle centered on the task of target classification. Because of the amount of data that needs to be processed and absorbed and because of the speed of the battle, fire control solutions on both sides of the battle must be automatic or at least computer aided. Tactical decision algorithms based on the probability of raid annihilation and other analyses were briefly described in Chapter 1 and in earlier sections of this chapter.

In an effort to describe the status of present-day EW the battle between the fleet and a single ASM has been investigated in this work. The example threat is radar guided autonomous ASM in its terminal phase of attacking the HVU. Prior to this phase, the ASM knows where the ship was and has information about its several physical characteristics.

This information may be from intelligence and/or testing coupled with the possibility that the ASM seeker has viewed the ship target earlier in its approach. For example, a modern sea-skimming ASM can travel 600 km on its clandestine approach at Mach 1. As it pops-up to view the scene it will sweep the sea surface where the ship was previously seen. Using a high PFA algorithm and an RCS window, a collection of potential targets will be seen in the angle swept range swath. If the ship has launched decoys, such as passive decoys like the Rubber Duck or active decoys (Nulka), multiple potential targets will be detected. Figure 7.17 illustrates the ASM, viewing a range swath containing the HVU, a single escort ship, and a single decoy.

The ASM will next utilize a set of one or more waveforms designed to enhance the difference between the HVU target, the smaller escort ship, and the false target features as described in the previous chapters. If the ASM changes waveforms it will need a few tens of milliseconds to flow data through its pipeline as described in Chapter 2. For example as described in Chapter 4, the several targets may be examined for RCS values and RCS Lag-1 value. The monopulse statistics may be examined for features associated with the

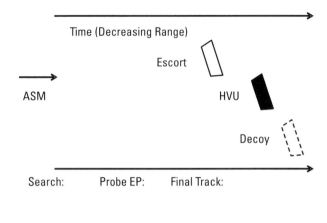

Figure 7.17 ASM attacking the HVU protected by a decoy.

extended nature of the true target. In addition, or as an option, while the ASM conducts a weave maneuver to confound kinetic weapon fire control systems the seeker can evaluate the Doppler characteristics as described in Chapter 5. Once the ASM chooses the highest probability HVU target it will tighten its track gates and switch to waveforms optimized for collecting precise guidance information.

As another option, the escort ship may generate false targets with the on-board EA system in an effort to confuse the scene of the ASM sensor. The ASM may switch to a very long pulse compression waveform with pulse to pulse random codes to make it impossible for the EA system to generate viable false targets within the range swath. Or the ASM may measure correlations of parameters between pairs of targets to sort the many targets into false targets and the true target. If the escort ship successfully generates a false target in an effort to seduce range and Doppler track gates the beat frequency detector (BFD) procedure will alert the ASM sensor. And of course while this work has dealt with EP for an LPI seeker, the seeker may have multiple sensors. Employing an IR sensor to detect the signature of a ship or passive radar to detect emanations from ship board radar can further confirm the conclusions from the LPI active radar sensor to properly classify the targets. The use of multiple active EA false targets is illustrated in Figure 7.18.

As described in Chapter 6, the escort ship (or the active decoy) may attempt to protect the fleet HVU by generating noise jamming to blind the

Adaptive EW

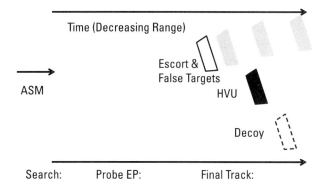

Figure 7.18 Escort ship active EA generated false targets.

active radar sensor as shown in Figure 7.19. The figure illustrates the location of the escort ship, the HVU, and the decoy, but assumes the escort ship has flooded the seeker active radar sensor with high-level noise radiation.

As the ASM begins its weave maneuver and detects this jamming, it can implement the two coherent sensors near-forward looking space-time adaptive

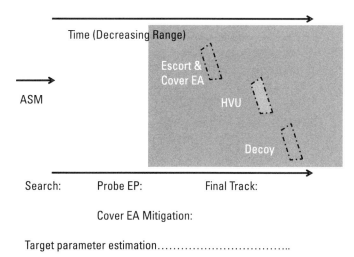

Figure 7.19 Escort ship active cover jamming EA.

processing algorithm to mitigate the cover jamming and reveal the ship locations and features. With prior information and observations, it is possible to obtain crude information about the possible ship location. This information combined with the observed data array is adequate to establish the suboptimal filters around each of the potential targets as described in Chapter 6. Using these filters, more detailed observations of the targets can be measured. This information can serve to assist guidance and at the same time improve the filters, bringing them closer to optimal prewhitening filters. This technique of adaptive filter estimation is a common means of suboptimal filtering exploited in many signal processing applications [3]. In this way the ASM can rapidly mitigate the jamming and ascertain the location of the HVU. This result is heuristically indicated in Figure 7.20.

7.4 Summary and Conclusions

The LLR mathematical formalism was described to illustrate a typical algorithm useful for automatic combat decision aid while incorporating the task of RTEAM with rapid sensor fusion. This exercise demonstrated how to model the threats, the weapons, and the sensors to achieve this capability. It was shown how to develop the sensor requirements and to evaluate the quantitative performance, including the performance range requirement. This approach illustrated the powerful result of being able to fuse information using scalar

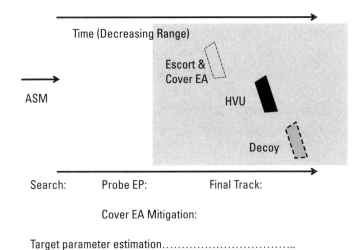

Figure 7.20 Escort ship active cover jamming EA mitigated by ASM EP.

arithmetic. It was indicated how to accumulate confidence in the results over time. This formalism has been previously applied to several field test exercises.

In a similar manner, several of the above EP techniques can be processed in parallel in the autonomous ASM sensor and standard statistical techniques used to weight and combine the several techniques. The modern ASM radar sensor may generate an LPI waveform such as a chirp or a long-phase coded waveform during its search mode that is optimized for target detection. With a large PFA, any and all targets in the angle and range swath will be detected. Using a combination of rudimentary EP techniques from Chapter 4, such as RCS level, Lag-1 statistics, and range/Doppler structure (e.g., length), one or more potential targets can be identified as true targets.

As the antenna is aimed at these detected targets, a different waveform (perhaps a probe waveform from Chapter 5) can be employed to better distinguish the correct target from the false targets and decoys. If there are multiple false targets, correlating parameters can rapidly identify the false targets. Use of the stepped frequency and fixed frequency waveforms can be used by the ASM sensor to deliberately probe the targets for physical features that are difficult to duplicate with present EA systems and concepts. The variance of the monopulse measurement and other DSP algorithms can be evaluated as the range decreases. If the EA systems attempt to blind the ASM sensor by high-power cover noise, the techniques in Chapter 6 can be employed to mitigate this EA and identify the proper Doppler range cell containing the HVU.

The result is an ASM threat that is very efficient at detecting and classifying the HVU in the presence of EA. As a counter to this capability, it is imperative that the EA techniques be improved to adequately match this technology advance by utilizing physically accurate techniques to mimic the characteristics of the HVU and provide a truly viable alternative to the true target. To do this task, the EA engineer must understand the DSP techniques being developed by the ASM engineer. The EA techniques must be improved and fully integrated with fleet defensive weapons and sensors to optimize the defenses and to add full real-time cognizance of the battle.

As stated at the beginning of this tract, the modern radar guided ASM employing optimal signal processing of the data in the two receivers is a formidable threat that can mitigate most of the standard present-day EA techniques. A main result of this work is that there are many target features and EP probe techniques available to provide the ASM sensor with an abundance of information to counter EA and to classify the desired target. The main observation is that to a large extent, any or all of these rapid and efficient EP algorithms can be implemented via software changes in modern and existing ASM seeker sensors equipped with high-speed digital signal processing capabilities.

References

[1] Schleher, D. C., *Electronic Warfare in the Information Age*, Norwood, MA: Artech House, 1999.

[2] Pace, P., *Detecting and Classifying Low Probability of Intercept Radar*, Norwood, MA: Artech House, 2009.

[3] Genova, J. J., *Non-Invasive Medical Monitor System*, Patent 5,590,650, Alexandria, VA, January 7, 1997.

About the Author

Dr. James Genova received his B.S. from Case Institute of Technology in 1968 and his M.A. and Ph.D. in theoretical elementary particle physics from the State University of New York at Stony Brook in 1971 and 1973, respectively. In his early career, Dr. Genova was a systems flight test engineer and then performed as a research scientist and project manager on various antisubmarine warfare (ASW) projects. The wide-band ambiguity function was the first successful application of coherent DSP to ocean acoustic data (sonobuoys, towed arrays, and SOSUS arrays). He continued with the application of coherent DSP to various unstable acoustic signals.

In 1978, Dr. Genova became project manager of a joint DARPA and NAVELEX R&D program investigating the application of feedback control to shipboard EW. He quickly realized the significant benefit to be realized from adaptive feedback control EW depends on the successful application of DSP to EW signals. There followed an intensive study of ASM, shipboard EW, and various EA techniques, especially dual coherent source (DCS) EA, including cross-eye, cross-pol, and double-cross. He became an internationally recognized expert in DCS monopulse radar deception. This research phase culminated in an Advanced Technology Demonstration of EA effectiveness assessment conducted on a U.S. Navy ship in 1990.

In 2002, he joined the Naval Research Laboratory (NRL) EW Division as a staff scientist and retired in 2012. During his NRL tenure he won three consecutive awards for the best formal report of the year. He proposed and conducted several innovative R&D projects for the Office of Naval Research, including the application of space-time adaptive processing to naval ASM seeker processing and an innovative DSP technique for onboard naval EA effectiveness assessment. This latter project transitioned to a Future Naval Capability Program in FY2015. Dr. Genova's primary task at NRL was to develop the first three coherent radar seeker hardware simulators for naval ASM research and testing. These have now been used for R&D and testing (including hardware-in-the-loop anechoic chamber tests and captive carry tests) for 10+ years. Dr. Genova conducted and supported these and many other EW tests over the years, including the U.S. Navy testing at each RIMPAC exercise with specific naval allies.

Intelligence in the area of ASM seekers is primarily based on intercepts of RF transmissions and conjectures about the RF processing. The DSP behind the RF receivers is totally software configurable and not directly observable. This gap in understanding leads to a significant flaw in naval EW testing philosophy. During official testing, simulators only implement what is confirmed by intelligence gathering, that is, none of the DSP algorithms. Previously, the EW battle was an effort to exploit flaws in the ASM seeker sensor and to attack the detection and/or localization functions. Modern LPI radar assets are hardened with respect to detection and localization. The present focus of the LPI ASM seeker development is on the target classification function via DSP, and the EW battle must now be directed at this classification function. Dr. Genova has researched the literature and other sources to discern what may be currently implemented in the ASM DSP. Using primarily the publications by Russian and PRC engineers directed at target classification DSP, Dr. Genova has implemented and tested a variety of these techniques via the NRL coherent radar seeker hardware simulators. While ASM coherent radar seekers are a formidable threat, if these and other ASM DSP techniques are implemented, standard naval EA systems are inadequate for their task. This book describes these ASM DSP algorithms presently ignored by EA engineers and the Intelligence community.

Index

Advanced Technology Demonstration (ATD), 120, 158, 232–233
Aerodynamic Radome, See Radome
Aircraft Carrier, See High Value Unit
Ambiguity Function, 172–173
Analog to Digital Converter (ADC)
    ADC, 32, 48–49, 74
    ADC Sampling, 52, 61
    Dynamic Range, 49, 216
    Saturation, 196, 216
Antenna
    ASM Antenna, 37–38
    ASM Antenna Types, 9, 11, 62
    Conical Scanning Antenna, 9–10
    Conical Scan on Receive Only, 10
    Flat Plate Dipole Array 9, 62–68, 100
    Gain Function, 63
    See also Antenna Beam Patterns
Antenna Beam Patterns
    Antenna Polarization, 33–34, 66–68, 100–104, 134–137, 158–159
    Beam Patterns, 62–68
    Parabolic Antenna Polarization, 66–68, 158–159
Anti-Ship Missile (ASM)
    ASM Modes, See Seeker
    ASM Seeker Sensor, See Seeker
    DF-21, 246
    Exocet, 5

Sea Skimming ASM, 12–13, 16, 19, 26, 81, 88, 243, 247
Styx Missile, 5
ASM Maneuver
    Pop Up Maneuver, 13, 82, 229, 247
    Weave or High g Maneuver, 164, 182, 217, 221, 225, 248–249
Automatic Gain Control (AGC), 8, 13, 46, 48–49, 76, 189
Azimuth Angle, 11, 66, 68, 81, 102, 147, 182

Beacon Rings, See Dual Coherent Source EA
Burn Through, 82, 94–96, 120

Carrier Frequency, 44, 85, 101, 130, 173–174
Cell Under Test (CUT), 193–194
Chaff, 4, 7, 14, 19, 26–28, 142–146, 180–181
Chirp Waveform, See Waveform LFM
Clutter
    Clutter to Noise Ratio (CNR), 90
    Sea Clutter, 88–90
Coherent Gain, See Digital Signal Processing Gain
Coherent Processing Interval (CPI), 68, 70, 74–76, 85–92

Constant False Alarm Rate (CFAR), See
    False Alarm Rate
Controller, 228–231
Corner Reflector, See Decoys
Correlation Function, 40–44, 205–207
    Lag 1 Correlation, 40–44, 133–141
    Normalized Correlation Function, 40
Cross Eye Jamming, See Dual Coherent
    Source Jamming
Cross Polarization Jamming, See Dual
    Coherent Source Jamming

Decoys
    Active Decoy, 117–140, 150
    Decoy False Target, 13–15, 17, 21,
        25–28, 109–110, 242–251
    Nulka, 14, 247
    Passive Decoy, 140–146, 151, 232–234
    Point Target, 149–151, 172–173, 175–
        181, 184
    Rubber Duck (UK manufactured
        FDS3), 140, 247
Detector, 46–48, 58
Detector Operating Characteristics
    (DOC), 239–240
DF-21, See Anti-Ship Missile
Digital RF Memory (DRFM), 79, 105–
    107, 116–124, 166–167, 175, 180
Digital Signal Processing Gain, 56–61,
    90, 95, 166–169, 196, 220
Dirac Delta Function, 35, 44
Direct Digital Synthesis (DDS), 179
Doppler
    Doppler Data Array, 69, 98, 113
    Doppler Processing, 75, 85, 138–139,
        174–185, 209
    Doppler Measurement, 71–75, 88–90,
        96, 103–104, 125–126
Doppler Beam Sharpening, 183–184
Doppler Extent, 143–44, 183–184
Doppler Range Array, 71–73, 109, 118,
    166, 182–184, 190

Doppler Resolution, 175
Double Cross Jamming, See Dual
    Coherent Source EA
Dual Coherent Source (DCS) EA
    Beacon Rings, 156
    Monopulse Angle Deception, 150–158
    Null Rings, 156–158
Dwell, 84–89
Dynamic Range, See Analog to
    Digital Converter

Electromagnetism
    Electromagnetic Pulse, 33–44
    Electric Field, 33–35, 44, 100, 106
    Transverse Wave, 33
Electronic Attack (EA)
    Angle Deception, 12, 117
    Blinking Jamming, 8
    Count Down Jamming, 8
    Cover Jamming
        Cover EA, 7, 28, 82, 106, 118–120
        Cover Jamming EP, 94–95, 110–
            112, 123, 188–222
    Keeper Pulse, 120, 232–233
    Low Duty Cycle, See Count
        Down Jamming
    Noise Jamming, See Cover Jamming
    Nonkinetic (NK) Weapon, 1–28
    Repeater EA, See Digital RF Memory
    Soft-Kill, See Nonkinetic Weapon
    See also Dual Coherent Source (DCS)
        EA
Electronic Protection (EP), 2, 112–114
Electronic Warfare (EW)
    EA Effectiveness Assessment, 244
    Real Time EA Effectiveness
        Assessment (RTEAM), 229–232
    See also Electronic Attack
    See also Electronic Protection
Elevation Angle, 11, 66, 159
Engagement Timeline, 229–233, 235,
    244

Exocet, See Anti-Ship Missile

False Alarm Rate (FAR). 76, 83–87, 91
Filter
    Band Pass Filter, 36, 48
    Low Pass Filter, 44, 46–48
    Matched Filter, 49–51, 58–60, 86–87, 165–166, 170
    Prewhitened Matched Filter, 200, 216
Fourier Transform (FT)
    Convolution, 39, 45, 53
    Discrete Fourier Transform (DFT), 52–55
    Zero Fill DFT, 55
Frequency, 33–37

Ghost Images, 138–139, 169
Guidance, See Navigation

High Value Unit (HVU), 4, 16–17, 27–28
Hypothesis Testing, 109–110, 194, 232–233

Image Processing, 80, 98
Impulse Function, See Dirac Delta Function
Information Warfare (IW)
    Deceptive Information Warfare, 3
    Hypothesis Testing, 109, 112–114
    Target Classification, 2, 13–19, 27, 31, 76, 81, 109
Intelligence
    Intelligence Gathering, 7, 15, 247, 254
    Intelligence Information, 17–18, 21, 122, 223, 229

Jamming, See Electronic Attack
Jamming to Noise Ratio (JNR), 194, 196
Jamming to Signal Ratio (JSR), 123, 197, 216, 218–222

Keeper Pulse, See Electronic Attack

Lag 1 Correlation, See Correlation
LFM Frequency Offset, See Ambiguity Function
Local Oscillator (LO), 36
Log Likelihood Ratio (LLR), 194–198, 215–216, 235–245

Marcum Target, 123, 128, 132
Maxwell, James Clerk, 33, 66, 68, 99
Miss Distance, 21, 229, 242–243
Missed Pulses False Target, See Ghost Images
Mixer, 37, 46, 48, 71
Monopulse
    Angle Measurement, 11–13, 66, 105, 108
    Monopulse EA, See Dual Coherent Source EA
    Monopulse EP, 137–138, 146–159, 189–199

Naval fleet, See Targets
Navigation
    ASM Guidance, 1, 2, 6, 8, 19
    Proportional Navigation, 9
Neyman-Pearson Criteria, 195, 237, 240
Noncoherent Gain, See Digital Signal Processing Gain
Nulka, See Decoys
Null Rings, See Dual Coherent Source Jamming

Observer, 228, 231
Over the Horizon Radar, 246
Otomat, See Anti-Ship Missile

Polarization
    Antenna Polarization, 62–68, 100, 158–159
    Electromagnetic, 34, 38

Polarization *(Cont.)*
  Jones Vector, 100, 206
  Polarization EP, 113, 134–137, 221
  Polarization Scatter Matrix, 102, 104
Power
  Average Power, 86–90
  Peak Power, 58, 86–88, 94, 99, 105, 165
  Power Spectrum, 40
  Total Power, 40
Probability
  A Posteriori Probability, 18, 21, 23, 26, 223, 242
  A Priori Probability, 18, 21, 23–25, 223, 229, 232–233
  Lethality Reduction, 24–25
  Probability of Kill, 18, 21–24, 233
  Probability of Raid Annihilation (PRA), 24–26
Pulse Repetition Frequency (PRF), 85
Pulse Repetition Interval (PRI), 85, 91–92
Pulse Width (PW), 38, 57–58, 61

Quadrature Detector, See Detector

Radar
  Coherent Processing Interval (CPI), 68–76, 85–92, 96–97, 106, 111–113, 166–169
  Coherent Radar, 16, 17, 19, 46–47, 56–59, 81–94
  Continuous Wave (CW) Radar, 38
  Frequency Bands, 36–37
  Low Probability of Intercept (LPI) Radar, 13, 17, 19, 49, 51, 58, 165
  Noncoherent Radar, 81–94
  Pulse, 7–9, 38–40, 45
  Pulsed Doppler Radar, See Low Probability of Intercept Radar
  Radar Horizon, 63, 81, 112, 229
  Radar Range, 45–46, 51
  Range Equation, 86, 94–96, 99, 106, 148
Radar Cross Section (RCS), 80, 86, 94, 99, 102, 122–138, 141
Radome, 19, 66–68, 100, 158–159
Range Extent, 98, 115–160
Range Resolution, 57–61, 88–89, 165–167, 172, 176–177
Range Swath, 68, 81–82, 84–94
Receiver Operating Characteristics (ROC), See Detector Operating Characteristic
Resolution, See Doppler Resolution and Range Resolution
Rubber Duck, See Decoys

Sample Rate, 42, 53, 236, 240–242
Satellite, 246
Scatter Matrix, 102–105
Schwartz Inequality, 200, 202
Seeker Modes
  Classification Mode, 81, 109
  Detection, See Search Mode
  Home On Jam Mode (HOJ), 82, 95, 118, 119, 123, 188, 190, 221, 228, 233
  Reacquisition Mode, See Search Mode
  Search Mode, 63, 73, 76, 81–84, 109, 113, 234, 251
  Surveillance Mode, See Search Mode
  Track Mode, 76, 82, 105, 109, 113, 117, 119–120, 124, 229–234
  Tracking Gates, 6–9, 27, 105, 120, 159, 227, 233
Seeker Sensor, 19
Sensor, alternative or multiple
  Electro Optical, 16
  IR, 13, 16, 18
  Radar, 14–15
Sensor Fusion, 241–242, 250
Signal to Interference Ratio (SIR), 193, 197–198, 200–203, 209, 213, 216–221

Signal to Noise Ratio (SNR), 50–51, 56–57, 87–94, 99, 148, 192, 196–221
Sinc Function, 38, 53, 65, 170, 175
Spectrum, 34–45, 52, 53, 88–89
Speed of Light, 33, 36
Steering Vector, 200–203, 209–221
Styx, See Anti-Ship Missile
Swarm Attack, 2, 26
Swerling Target, 130–131

Targets
  Aircraft Carrier, See High Value Unit
  Decoy, See Decoys
  Doppler Extent, See Doppler Extent
  Escort Ship, 16, 19, 26, 118, 246–250
  Extended Target, 115–160
  False Target, See Digital RF Memory
  Marcum Target, See Marcum Target
  Swerling Target, See Swerling Target
  Target Classification and Target Features, See Seeker Modes
Terrain Bounce EA, See Dual Coherent Source EA
Threat, See Anti-Ship Missile

Waveform
  Barker Code, 59–61, 165–168
  Chirp Waveform, See Ambiguity Function
  Phase Code, 58, 165–167, 251
  Probe Waveform, 173, 181–185
  Pulse Compression, 58–62, 163, 165–173, 185
  Random Code, 60, 164–169, 185, 248
  Stepped Frequency Waveform, 173–181
Wavelength, 36, 128–130
Weapon
  Close In Weapons System (CIWS), 5, 16, 17, 246
  Hard-Kill, See Kinetic Weapons
  High Energy Laser Beam, 5, 246
  Kinetic Weapon, 1, 5, 6, 7, 16, 18, 22–26, 63, 164, 233
  Nonkinetic Weapon (NK), 1, 6, 16–28
  Soft-Kill, See Nonkinetic Weapons
  See also Electronic Attack (EA)
  See also Anti-Ship Missile
Weighting Function, 39, 59, 65
Window Function, 38–40, 45, 53, 64–66, 70

The Artech House Electronic Warfare Library

Activity-Based Intelligence: Principles and Applications,
Patrick Biltgen and Stephen Ryan

Advances in Statistical Multisource-Multitarget Information Fusion,
Ronald P. S. Mahler

Antenna Systems and Electronic Warfare Applications,
Richard A. Poisel

Electronic Intelligence: The Analysis of Radar Signals, Second Edition, Richard G. Wiley

Electronic Warfare for the Digitized Battlefield, Michael R. Frater and Michael Ryan

Electronic Warfare in the Information Age, D. Curtis Schleher

Electronic Warfare Receivers and Receiving Systems,
Richard A. Poisel

Electronic Warfare Signal Processing, James Genova

Electronic Warfare Target Location Methods, Richard A. Poisel

EW 101: A First Course in Electronic Warfare, David L. Adamy

EW 103: Tactical Battlefield Communications Electronic Warfare,
David L. Adamy

EW 104: EW Against a New Generation of Threats, David L. Adamy

Foundations of Communications Electronic Warfare,
Richard A. Poisel

High-Level Data Fusion, Subrata Das

Human-Centered Information Fusion, David L. Hall and
John M. Jordan

Information Warfare and Organizational Decision-Making,
Alexander Kott, editor

Information Warfare and Electronic Warfare Systems,
Richard A. Poisel

Information Warfare Principles and Operations, Edward Waltz

Introduction to Communication Electronic Warfare Systems, Richard A. Poisel

Knowledge Management in the Intelligence Enterprise, Edward Waltz

Mathematical Techniques in Multisensor Data Fusion, Second Edition, David L. Hall and Sonya A. H. McMullen

Modern Communications Jamming Principles and Techniques, Richard A. Poisel

Principles of Data Fusion Automation, Richard T. Antony

Stratagem: Deception and Surprise in War, Barton Whaley

Statistical Multisource-Multitarget Information Fusion, Ronald P. S. Mahler

Tactical Communications for the Digitized Battlefield, Michael Ryan and Michael R. Frater

Target Acquisition in Communication Electronic Warfare Systems, Richard A. Poisel

For further information on these and other Artech House titles, including previously considered out-of-print books now available through our In-Print-Forever® (IPF®) program, contact:

Artech House	Artech House
685 Canton Street	16 Sussex Street
Norwood, MA 02062	London SW1V 4RW UK
Phone: 781-769-9750	Phone: +44 (0)20-7596-8750
Fax: 781-769-6334	Fax: +44 (0)20-7630-0166
e-mail: artech@artechhouse.com	e-mail: artech-uk@artechhouse.com

Find us on the World Wide Web at: www.artechhouse.com